Silicon Carbide: Material Growth, Device Processing and Applications

Silicon Carbide: Material Growth, Device Processing and Applications

Marilena Vivona
Mike Jennings

MDPI

Basel • Beijing • Wuhan • Barcelona • Belgrade • Novi Sad • Cluj • Manchester

Editors

Marilena Vivona
Consiglio Nazionale
delle Ricerche
Istituto per la Microelettronica
e Microsistemi (CNR-IMM)
Catania
Italy

Mike Jennings
Centre for Integrative
Semiconductor Materials
(CISM)
Swansea University
Swansea
United Kingdom

Editorial Office
MDPI AG
Grosspeteranlage 5
4052 Basel, Switzerland

This is a reprint of articles from the Special Issue published online in the open access journal *Materials* (ISSN 1996-1944) (available at: www.mdpi.com/journal/materials/special_issues/Silicon_Carbide_Mater_Growth_Device_Process_Appl).

For citation purposes, cite each article independently as indicated on the article page online and as indicated below:

Lastname, A.A.; Lastname, B.B. Article Title. *Journal Name* **Year**, *Volume Number*, Page Range.

ISBN 978-3-7258-2238-6 (Hbk)
ISBN 978-3-7258-2237-9 (PDF)
doi.org/10.3390/books978-3-7258-2237-9

Contents

Marilena Vivona and Mike Jennings
Silicon Carbide: Material Growth, Device Processing, and Applications
Reprinted from: *Materials* **2024**, *17*, 4571, doi:10.3390/ma17184571 . . . . **1**

Laura Anoldo, Edoardo Zanetti, Walter Coco, Alfio Russo, Patrick Fiorenza and Fabrizio Roccaforte
4H-SiC MOSFET Threshold Voltage Instability Evaluated via Pulsed High-Temperature Reverse Bias and Negative Gate Bias Stresses
Reprinted from: *Materials* **2024**, *17*, 1908, doi:10.3390/ma17081908 . . . . **3**

Jiashu Qian, Limeng Shi, Michael Jin, Monikuntala Bhattacharya, Atsushi Shimbori and Hengyu Yu et al.
Modeling of Charge-to-Breakdown with an Electron Trapping Model for Analysis of Thermal Gate Oxide Failure Mechanism in SiC Power MOSFETs
Reprinted from: *Materials* **2024**, *17*, 1455, doi:10.3390/ma17071455 . . . . **15**

Suman Das, Yongju Zheng, Ayayi Ahyi, Marcelo A. Kuroda and Sarit Dhar
Study of Carrier Mobilities in 4H-SiC MOSFETS Using Hall Analysis
Reprinted from: *Materials* **2022**, *15*, 6736, doi:10.3390/ma15196736 . . . . **31**

Shengnan Zhu, Tianshi Liu, Junchong Fan, Hema Lata Rao Maddi, Marvin H. White and Anant K. Agarwal
Effects of JFET Region Design and Gate Oxide Thickness on the Static and Dynamic Performance of 650 V SiC Planar Power MOSFETs
Reprinted from: **2022**, *15*, , doi:10.3390/ma15175995 . . . . **41**

Razvan Pascu, Gheorghe Pristavu, Dan-Theodor Oneata, Gheorghe Brezeanu, Cosmin Romanitan and Nikolay Djourelov et al.
Thorough Wide-Temperature-Range Analysis of Pt/SiC and Cr/SiC Schottky Contact Non-Uniformity
Reprinted from: *Materials* **2024**, *17*, 400, doi:10.3390/ma17020400 . . . . **52**

Seul-Ki Kim, Hajun Kim, Hyun Sik Kim, Tae Eun Hong, Younki Lee and Eun Young Jung
Characteristics of the Discoloration Switching Phenomenon of 4H-SiC Single Crystals Grown by PVT Method Using ToF-SIMS and Micro-Raman Analysis
Reprinted from: *Materials* **2024**, *17*, 1005, doi:10.3390/ma17051005 . . . . **65**

Christina Villeneuve-Faure, Abdelhaq Boumaarouf, Vishal Shah, Peter M. Gammon, Ulrike Lüders and Rosine Coq Germanicus
SiC Doping Impact during Conducting AFM under Ambient Atmosphere
Reprinted from: *Materials* **2023**, *16*, 5401, doi:10.3390/ma16155401 . . . . **79**

Dalal Fadil, Wlodek Strupinski, Emiliano Pallecchi and Henri Happy
Analysis of Local Properties and Performance of Bilayer Epitaxial Graphene Field Effect Transistors on SiC
Reprinted from: *Materials* **2024**, *17*, 3553, doi:10.3390/ma17143553 . . . . **94**

Emanuela Schilirò, Patrick Fiorenza, Raffaella Lo Nigro, Bruno Galizia, Giuseppe Greco and Salvatore Di Franco et al.
Al_2O_3 Layers Grown by Atomic Layer Deposition as Gate Insulator in 3C-SiC MOS Devices
Reprinted from: *Materials* **2023**, *16*, 5638, doi:10.3390/ma16165638 . . . . **106**

Shengnan Zhu, Tianshi Liu, Junchong Fan, Arash Salemi, Marvin H. White and David Sheridan et al.
A New Cell Topology for 4H-SiC Planar Power MOSFETs for High-Frequency Switching
Reprinted from: *Materials* **2022**, *15*, 6690, doi:10.3390/ma15196690 . . . . **116**

Haiyan Ou, Xiaodong Shi, Yaoqin Lu, Manuel Kollmuss, Johannes Steiner and Vincent Tabouret et al.
Novel Photonic Applications of Silicon Carbide
Reprinted from: *Materials* **2023**, *16*, 1014, doi:10.3390/ma16031014 . . . . **122**

materials MDPI

Editorial

Silicon Carbide: Material Growth, Device Processing, and Applications

Marilena Vivona [1,*] and Mike Jennings [2]

1 Institute for Microelectronics and Microsystems, National Research Council of Italy, 95121 Catania, Italy
2 Centre for Integrative Semiconductor Materials (CISM), Swansea University, Swansea SA1 8EN, UK; m.r.jennings@swansea.ac.uk
* Correspondence: marilena.vivona@imm.cnr.it

The continuous demand for electronic devices operating at increasing current and power levels, as well as at high temperatures and in harsh environments, has driven research into wide-band gap (WBG) semiconductors over the last three decades. This is because, due to their outstanding physical properties, WBG semiconductors can overcome the physical and electrical limits imposed by the use of conventional silicon devices [1]. Among WBG materials, the 4H hexagonal polytype of silicon carbide (4H-SiC) is the most promising for use in power electronics applications in the medium-to-high voltage range (600–3000 V) [2,3]. However, to achieve optimized performances with these 4H-SiC devices, a full understanding of the fundamental material properties, processing technologies, and carrier transport mechanisms associated with this semiconductor material is required. In this sense, there is still plenty of room for the progress of scientific and technological research related to this material. On the one hand, an improvement to the performance of existing power devices in terms of efficiency and reliability is a key aim; on the other hand, the uses of 4H-SiC could be extended to new cutting-edge technologies, e.g., quantum technologies and sensors.

This Special Issue, entitled "Silicon Carbide: Material Growth, Device Processing, and Applications", showcases a collection of papers on technological developments in SiC-based devices, including ten original research articles and one review paper.

The topics addressed in the Special Issue can be categorized into three main themes: (1) investigations into the fundamental characteristics of conventional 4H-SiC devices, (2) suggestions of new approaches to developing improved devices, and (3) the use of SiC devices in emerging technology fields, such as quantum technology applications.

Among the papers focusing on important aspects of conventional 4H-SiC-based devices, methods for evaluating the reliability of the critical SiO_2/SiC interface in planar MOSFET devices are discussed in the determination of real stress effects under extreme operational conditions [4], as well as in the assessment of the charge-to-breakdown in thermal gate SiO_2 [5]. The effects of the various scattering mechanisms on the channel conduction are also investigated in 4H-SiC MOSFETs [6], whilethe performance of 4H-SiC MOSFETS has been evaluated using various JFET and gate oxide process parameters [7]. The inhomogeneity in Schottky barrier diodes (Pt/4H-SiC and Cr/4H-SiC contacts) has also been discussed according to the parallel-diode model and evaluated under the conditions of a wide range of operational temperatures and biases [8].

A second group of papers focuses on the issues observed during standard approaches to SiC material growth and characterization. Notably, the discoloration switching phenomenon seen during a single crystal grown of 4H-SiC [9] is discussed, as well as the direct anodic oxidation phenomenon observed on the SiC surface during conductive AFM (C-AFM) measurements in an ambient atmosphere [10].

Regarding possible advancements in SiC devices via the proposal of innovative processing solutions, a 2D material (bilayer epitaxial graphene) has been investigated and evaluated for the fabrication of new field-effect transistors on SiC [11], while a high-permittivity

Citation: Vivona, M.; Jennings, M. Silicon Carbide: Material Growth, Device Processing, and Applications. *Materials* **2024**, *17*, 4571. https://doi.org/10.3390/ma17184571

Received: 29 August 2024
Accepted: 9 September 2024
Published: 18 September 2024

dielectric Al_2O_3/SiO_2 stack is proposed as a gate oxide for novel devices based on the cubic polytype of SiC (3C-SiC) [12]. Additionally, new possible designs are presented for cell topologies for 4H-SiC Planar Power MOSFETs for high-frequency power applications [13].

Finally, 4H-SiC has also been discussed as an emerging material in the photonics field. A review paper dedicated to the silicon carbide on insulator stack (SiCOI) [14] provides an interesting roadmap for further developments in the use of the SiCOI key structure in quantum photonic integrated circuit applications.

Of course, due to the broadness of 4H-SiC technology, the present collection cannot provide a comprehensive presentation of all the issues. However, we are confident that fundamental properties and interesting approaches have been presented and discussed in these papers.

Acknowledgments: We, the Guest Editors, thank all contributing authors for submitting their scientific results, as well as the peer reviewers, whose attentive efforts and expert knowledge made the preparation and publication of this Special Issue possible. We hope that this Special Issue will offer interesting inputs toward advancements in 4H-SiC-based technology.

Conflicts of Interest: The authors declare no conflicts of interest.

References

1. Ren, F.; Zopler, J.C. *Wide Band Gap Electronic Devices*; World Scientific Publishing: Singapore, 2003.
2. She, X.; Huang, A.Q.; Lucia, O.; Ozpineci, B. Review of Silicon Carbide Power Devices and Their Applications. *IEEE Trans. Ind. Electron.* **2017**, *64*, 8193–8205. [CrossRef]
3. Kimoto, T.; Cooper, J.A. *Fundamentals of Silicon Carbide Technology*; John Wiley & Sons: Singapore, 2014.
4. Anoldo, L.; Zanetti, E.; Coco, W.; Russo, A.; Fiorenza, P.; Roccaforte, F. 4H-SiC MOSFET Threshold Voltage Instability Evaluated via Pulsed High-Temperature Reverse Bias and Negative Gate Bias Stresses. *Materials* **2024**, *17*, 1908. [CrossRef] [PubMed]
5. Qian, J.; Shi, L.; Jin, M.; Bhattacharya, M.; Shimbori, A.; Yu, H.; Houshmand, S.; White, M.H.; Agarwal, A.K. Modeling of Charge-to-Breakdown with an Electron Trapping Model for Analysis of Thermal Gate Oxide Failure Mechanism in SiC Power MOSFETs. *Materials* **2024**, *17*, 1455. [CrossRef] [PubMed]
6. Das, S.; Zheng, Y.; Ahyi, A.; Kuroda, M.A.; Dhar, S. Study of Carrier Mobilities in 4H-SiC MOSFETS Using Hall Analysis. *Materials* **2022**, *15*, 6736. [CrossRef] [PubMed]
7. Zhu, S.; Liu, T.; Fan, J.; Maddi, H.L.R.; White, M.H.; Agarwal, A.K. Effects of JFET Region Design and Gate Oxide Thickness on the Static and Dynamic Performance of 650 V SiC Planar Power MOSFETs. *Materials* **2022**, *15*, 5995. [CrossRef] [PubMed]
8. Pascu, R.; Pristavu, G.; Oneata, D.-T.; Brezeanu, G.; Romanitan, C.; Djourelov, N.; Enache, A.; Draghici, F.; Ivan, A.M.; Ceuca, E. Thorough Wide-Temperature-Range Analysis of Pt/SiC and Cr/SiC Schottky Contact Non-Uniformity. *Materials* **2024**, *17*, 400. [CrossRef] [PubMed]
9. Kim, S.-K.; Kim, H.; Kim, H.S.; Hong, T.E.; Lee, Y.; Jung, E.Y. Characteristics of the Discoloration Switching Phenomenon of 4H-SiC Single Crystals Grown by PVT Method Using ToF-SIMS and Micro-Raman Analysis. *Materials* **2024**, *17*, 1005. [CrossRef] [PubMed]
10. Villeneuve-Faure, C.; Boumaarouf, A.; Shah, V.; Gammon, P.M.; Lüders, U.; Coq Germanicus, R. SiC Doping Impact during Conducting AFM under Ambient Atmosphere. *Materials* **2023**, *16*, 5401. [CrossRef] [PubMed]
11. Fadil, D.; Strupinski, W.; Pallecchi, E.; Happy, H. Analysis of Local Properties and Performance of Bilayer Epitaxial Graphene Field Effect Transistors on SiC. *Materials* **2024**, *17*, 3553. [CrossRef] [PubMed]
12. Schilirò, E.; Fiorenza, P.; Nigro, R.L.; Galizia, B.; Greco, G.; Di Franco, S.; Bongiorno, C.; La Via, F.; Giannazzo, F.; Roccaforte, F. Al_2O_3 Layers Grown by Atomic Layer Deposition as Gate Insulator in 3C-SiC MOS Devices. *Materials* **2023**, *16*, 5638. [CrossRef] [PubMed]
13. Zhu, S.; Liu, T.; Fan, J.; Salemi, A.; White, M.H.; Sheridan, D.; Agarwal, A.K. A New Cell Topology for 4H-SiC Planar Power MOSFETs for High-Frequency Switching. *Materials* **2022**, *15*, 6690. [CrossRef] [PubMed]
14. Ou, H.; Shi, X.; Lu, Y.; Kollmuss, M.; Steiner, J.; Tabouret, V.; Syväjärvi, M.; Wellmann, P.; Chaussende, D. Novel Photonic Applications of Silicon Carbide. *Materials* **2023**, *16*, 1014. [CrossRef] [PubMed]

materials

MDPI

Article

4H-SiC MOSFET Threshold Voltage Instability Evaluated via Pulsed High-Temperature Reverse Bias and Negative Gate Bias Stresses

Laura Anoldo [1,*], Edoardo Zanetti [1], Walter Coco [1], Alfio Russo [1], Patrick Fiorenza [2,*] and Fabrizio Roccaforte [2]

[1] STMicroelectronics, Stradale Primosole, 50, 95125 Catania, Italy
[2] Consiglio Nazionale delle Ricerche-Istituto per la Microelettronica e Microsistemi (CNR-IMM), Strada VIII 5, 95121 Catania, Italy
* Correspondence: laura.anoldo@st.com (L.A.); patrick.fiorenza@cnr.it (P.F.)

Abstract: This paper presents a reliability study of a conventional 650 V SiC planar MOSFET subjected to pulsed HTRB (High-Temperature Reverse Bias) stress and negative HTGB (High-Temperature Gate Bias) stress defined by a TCAD static simulation showing the electric field distribution across the SiC/SiO_2 interface. The instability of several electrical parameters was monitored and their drift analyses were investigated. Moreover, the shift of the onset of the Fowler–Nordheim gate injection current under stress conditions provided a reliable method to quantify the trapped charge inside the gate oxide bulk, and it allowed us to determine the real stress conditions. Moreover, it has been demonstrated from the cross-correlation, the TCAD simulation, and the experimental ΔV_{th} and ΔV_{FN} variation that HTGB stress is more severe compared to HTRB. In fact, HTGB showed a 15% variation in both ΔV_{th} and ΔV_{FN}, while HTRB showed only a 4% variation in both ΔV_{th} and ΔV_{FN}. The physical explanation was attributed to the accelerated degradation of the gate insulator in proximity to the source region under HTGB configuration.

Keywords: pMOSFET; 4H-SiC; HTGB; HTRB; pulsed stress; threshold voltage instability

Citation: Anoldo, L.; Zanetti, E.; Coco, W.; Russo, A.; Fiorenza, P.; Roccaforte, F. 4H-SiC MOSFET Threshold Voltage Instability Evaluated via Pulsed High-Temperature Reverse Bias and Negative Gate Bias Stresses. *Materials* **2024**, *17*, 1908. https://doi.org/10.3390/ma17081908

Academic Editor: Abderrahim Yassar

Received: 22 March 2024
Revised: 11 April 2024
Accepted: 17 April 2024
Published: 20 April 2024

1. Introduction

Due to its superior physical properties compared to silicon (Si), such as a wide bandgap of about 3.2 eV, a critical electric field, and high thermal conductivity, Silicon carbide (4H-SiC), which is a wide-band gap (WBG) semiconductor material, is becoming the emerging candidate for automotive applications, making power transistors with a high power density module featuring high blocking voltage and ultra-low conduction resistance [1–5]. In this context, High-Temperature Reverse Bias (HTRB) and High-Temperature Gate Bias (HTGB) are routinely performed product qualification tests for PowerMOSFET manufacturing [6–14]. In particular, the HTGB test is designed to electrically stress the gate insulator by applying a DC bias voltage at a high temperature, detecting any drift in electrical parameters caused by a significant number of charge traps at and near the SiC/SiO_2 interface, and bulk oxide traps generated as the oxide has been deposed. The HTRB test aims to monitor the drain terminal leakage current of the devices under High-Temperature Reverse Bias conditions over a period of time. HTRB combines electrical and thermal stress; this test can be used to check junction integrity, crystal defects, and ionic-contamination level, which can reveal device weaknesses or degradation effects in the field depletion structures at device edge termination and in surface passivation. The aim of this work is to compare the effects of the, namely, similar—in terms of the electric field across the insulator layer—negative gate bias stress (HTGB) and drain bias stress (HTRB) on PowerMOSFET, respectively. In this context, TCAD simulations were used to select the experimental conditions to obtain a fair comparison of the two methods. In fact, a comparison of the results obtained on PowerMOSFET aging under both HTRB and HTGB, leading to a different drift or degradation of the threshold voltage (V_{th}) and to the onset of the Fowler–Nordheim gate conduction (V_{FN}), can be fundamental to refine the electrical characterization protocols for device manufacturers [15–21]. This comparison is of fundamental relevance for device manufacturing and to define the route to "zero-failure" of the field of device under its "mission-profile".

2. Materials and Methods

Several 4H-SiC wafers containing vertical PowerMOSFET transistors were investigated in this work. MOSFETs are fabricated on 150 mm wafers on an n-type (0001) 4H-SiC 180 μm thick substrate with a nitrogen-doping concentration of 4–5 $\times$ 10^{18} cm^{-3} and resistivity of 0.012–0.025 Ω·cm. The n-type drift epitaxial layer is 12 μm thick with a nitrogen-doping concentration of 8 $\times$ 10^{15} cm^{-3} and is grown by chemical vapor deposition (CVD) in a warm wall multi-wafer reactor. The gate insulation layer is a 50 nm thick deposited oxide layer through a Low-Pressure Chemical Vapor Deposition (LPCVD) process followed by a state-of-the-art NO-based post-oxide deposition annealing [22]. The semiconductor materials were characterized at the beginning of the device fabrication process using microscopic techniques in order to select those devices without any visible epitaxial defect that may affect the device's reliability [12,23]. Firstly, the optical inspection at the surface of the epitaxial layer is carried out with Candela 8520 using a KLA-Tencor equipment scan which allows us to detect surface defects such as droplets, carrots, triangles, micro-pits, etc. Another optical inspection is performed, using a KLA Altair inspection microscope, after the first photo-lithography step, which defines the pitch of the device, that is, the width, distance, and PowerMOSFET edges. Then, an accurate selection of good devices normally intended for sale is carried out, superimposing defectivity maps, shown in Figure 1, and EWS (Electrical Wafer Sorting). In particular, the EWS is used to select devices distributed on the whole wafer surface, as schematically described in Figure 1 by red squares avoiding "dotted" devices spotted as defect-containing devices by optical inspection performed before and during device fabrication processing.

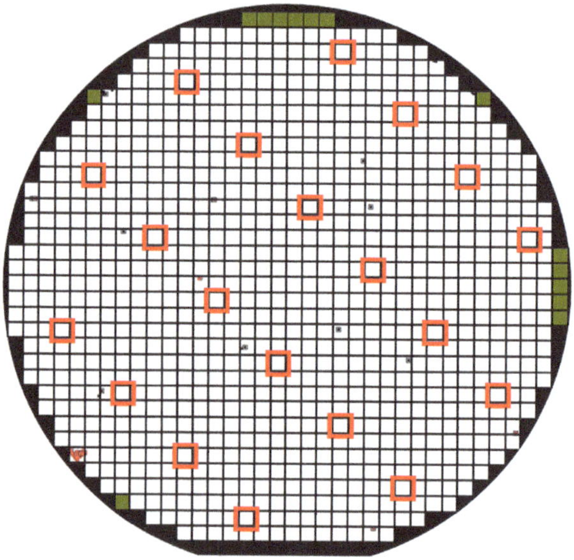

Figure 1. Defectivity maps at epitaxial inspection level. Red: The tested devices that are defect-free and good in the final test.

3. Device and Stress Procedure Description

A preliminary electrical characterization of the gate oxide and the channel leakage current under reverse gate bias was performed, and it aimed to obtain the onset of the Fowler–Nordheim (FN) tunneling conduction and to define input elements for the TCAD numerical simulation. This iteration is used to define the stress conditions to keep the "real oxide electric field" and the insulator similar, as much as possible, in both HTGB and HTRB conditions. In fact, FN tunneling current it is used to univocally estimate the oxide electric field, the effective thickness and the active semiconductor doping concentration [24–26]. Furthermore, both HTGB and HTRB conditions induce different electric field distributions across the MOSFET structure and they can be complementary used to understand which are the more robust and weak components of the elementary cell. In fact, as has been demonstrated previously by Fiorenza et al. [27], Fowler–Nordheim gate conduction can be used as reliable feedback on the control of the oxide electric field. In particular, Figure 2 shows the gate current characteristics (I_G-V_G) repeated several times, both on the positive and negative polarization region, and used to define the steady onset of the Fowler–Nordheim (FN) injection current and to determine gate current onset and effective oxide thickness [19,25,28]. Figure 2 shows the I_G-V_G sweeps collected twice consecutively, both on positive and negative polarizations. It can be noticed that in the negative branch of the I_G-V_G (Figure 2), the two consecutive curves do not overlap after repeated negative bias sweeps. This has been demonstrated previously [24] as a manifestation of transient trapping phenomena occurring at the SiO_2/4H-SiC gate system, resulting in an uncontrolled transient oxide field variation. These measurements are fundamental for giving feedback to the TCAD simulation to set the stress conditions in a well-known oxide field regime.

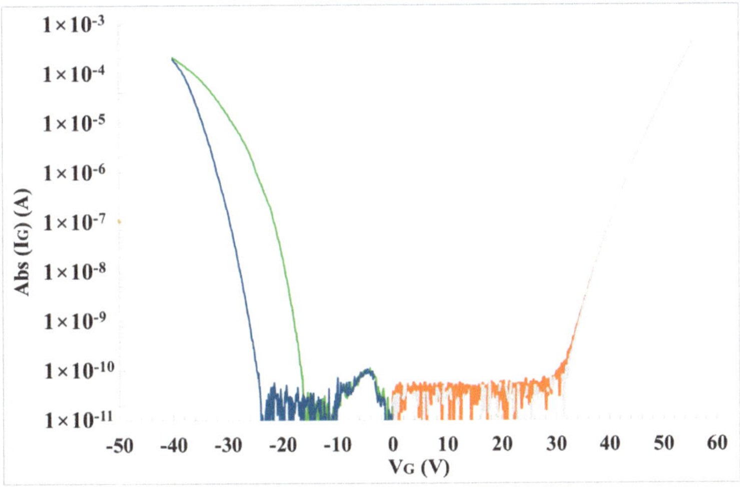

Figure 2. I_G-V_G characteristics of a 500 Å gate oxide on 650 V SiC PowerMOSFET under two consecutive positive (orange and grey) and negative (green and blue) bias sweeps at 200 °C.

On the other hand, Figure 3 shows the I_{DS}-V_{DS} characteristics used to determine the operation avalanche breakdown voltage (BV_{DSs}) of the device, the semiconductor parameters and to tune the HTRB stress conditions. The occurrence of the avalanche breakdown at a given BV_{DSs} value is fundamental in order to give feedback to the TCAD simulation to set the stress conditions in a well-known semiconductor field regime.

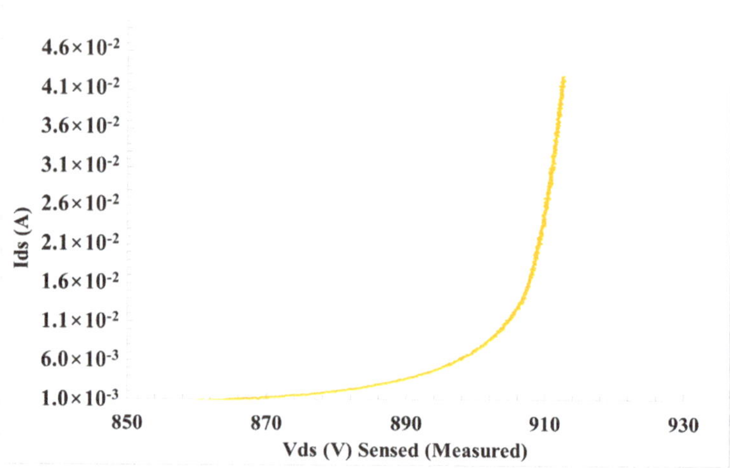

Figure 3. I_D-V_D characteristics under off conditions on a 650 V SiC PowerMOSFET at 200 °C.

As mentioned in the introduction, the electrical characterization, provided in Figures 2 and 3, was used to understand the experimental conditions in the negative polarization of the gate, such as negative gate bias stress (HTGB) and drain bias stress (HTRB), respectively. TCAD simulations were used to fine-tune the values of V_G and V_{DS} in order to obtain similar electric fields in the gate oxide (Figure 4). The schematic of the half-cell structure of SiC Planar MOSFET used in the numerical simulation is presented in Figure 4A. In particular, the static 2D TCAD physical simulations of the gate oxide electric field (E) distribution near the SiC-SiO_2 interface, under negative HTGB and HTRB conditions, of the 4H-SiC power MOSFET are also shown in Figures 4B and 4C, respectively. Finally, Figure 4D shows the scanning electron microscopy (SEM) performed on the MOSFET elementary cell structure. As can be seen, all the device components (source, body, JFET, etc.) are clearly distinguishable.

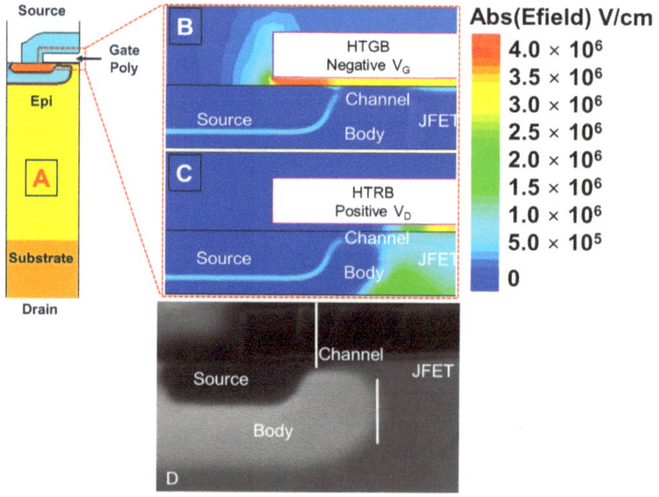

Figure 4. The half-cell structure of PowerMOSFET (**A**). Simulation of the gate oxide field under negative HTGB (**B**) and under HTRB (**C**) conditions. Scanning electron microscopy cross-section image of the MOSFET cell (**D**).

A simulation related to HTRB configuration, under the BV_{DSs} condition and limited to I_{DSs} = 1 mA, shows an oxide field maximum value in the JFET region (Figure 4C). On the other hand, the simulation under negative HTGB (Figure 4B), at an equivalent gate oxide field, shows a larger oxide field in the source region compared to the JFET. Hence, although the absolute value of the electric field induced on the gate oxide is the same, under negative HTGB, the field lines at the source/body interface are much higher than in the drift region due to the mismatch in the dielectric constant of the oxide and the semiconductor. Opposite behavior is seen in the case of the HTRB test, where the maximum electric field occurs in the drift region and is practically zero in the source/body region. In this context, it is mandatory to understand the device physics involved in both HTRB and HTGB configurations in order to clarify which particular phenomenon is involved in the different parts of the device degradation. In this scenario, it is fundamental to focus on the impact of these two different stress conditions (having a similar gate oxide field conditions), paying particular attention to the drift of key parameters, represented in Figure 5: the gate and drain terminals' leakage currents (I_{GSs} and I_{DSs}); the gate oxide conduction (V_{FN+} and V_{FN-} namely the Fowler–Nordheim voltage onset by electrons and hole currents coming from the semiconductor); the drain-source breakdown voltage (BV_{DSs}); the transistor threshold (V_{th}); the threshold with source and drain grounded (V_{tSAT}); threshold voltage at 10 µA (V_{tC}); the V_{DSs} voltage under conduction (ON-state) condition (V_{DSON}); and the body-drain diode forward bias (V_{FEC}). The testing protocol is performed using the Keysight Easy Expert platform, as schematically shown in Figure 5.

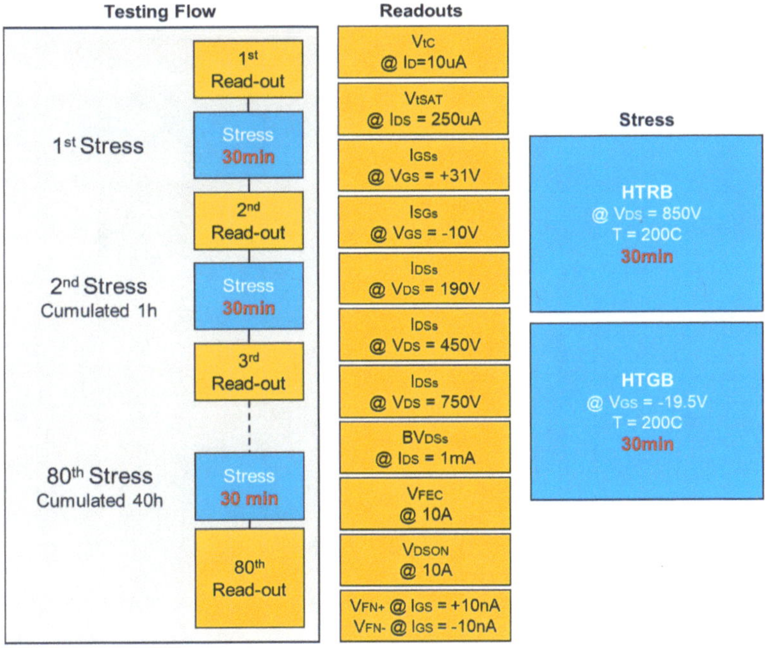

Figure 5. Stress cycling procedure.

The characterization flow—schematically depicted in Figure 5—consists of a sequence of stresses and read-outs, i.e., up to 40 h of cumulative stress which equates to 80 stress/read-outs. In each readout, the characteristic parameters of the MOSFET are measured; after the stresses sequence, either HTGB or HTRB are measured. Furthermore, the readout sequence is chosen from the mild perturbation to the more severe perturbation condition.

The investigation on the drift of the mentioned electrical parameters aims to understand which are the most sensitive parameters to drift under stress conditions. In order to induce significant drift in the selected parameters during stress, the measure/stress cycling routine allows us to control the Keysight B1505 device analyzer (Keysight, Santa Rosa, CA, USA) and the HV MPI-TS2000DP semiautomatic probe-station (MPI Corporation, Zhubei City, Taiwan) to perform wafer level measurements at a temperature of 200 °C. To take advantage of the simulation shown in Figure 4, and to accelerate device degradation, a higher V_{DSs} condition than BV_{DSs} (i.e. V_{DSs} = 890 V/I_{DSs} ~2 mA) for HTRB stress has been chosen, beside an equivalent V_{GSs} stress condition, in terms of a gate oxide field for negative HTBG at $V_{GSs} = -19.5$ V. The purpose is to put the transistor under an avalanche condition in order to accelerate the stress effects within 40 h. This particular high-voltage/high-current stress-sensing configuration needed an unconventional experimental setup that is worth describing. The enhancement of the power capability until a voltage of ~960 V and leakage current of ~33 mA are reached under pulsed mode [29] is allowed by a Keysight N1266 HVSMU current expander, with interconnection of the so-called Keysight SMU as the MC (Medium Current) and MP (Medium Power) of the B1505. Moreover, in order to switch from a high-voltage to a high-current test, a N1258A Module selector is implemented, as shown in the complete schematic of the connections in Figure 6.

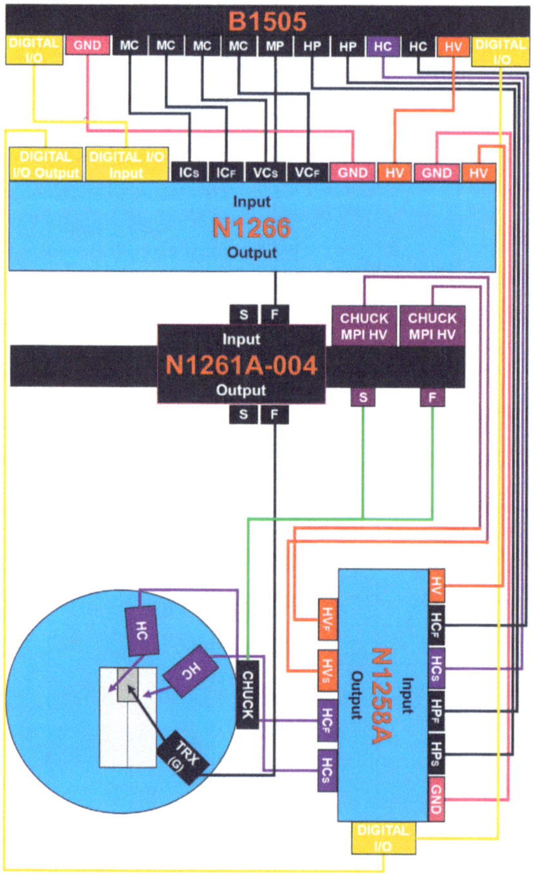

Figure 6. Test setup and system configuration schematic.

On the other hand, pulsed HTGB and HTRB stress occur close to the operative condition of the real application of the PowerMOSFET subjected to on/off hard switching. Hence, pulsed stress characterization was achieved by placing an oscilloscope with an

HV probe in series with the output of the resources. Although the N1266 module allows us to impose a maximum impulse of 1 ms, the resistance of the wiring system, and the consequent misalignment measured between forced voltage and applied voltage, led us to reduce the impulse to 500 µs. Thus, each stress cycle consisted of 1000 pulses with a 5 ms period, repeated 360 times for a total stress time of 1800 s, as shown in Figure 7.

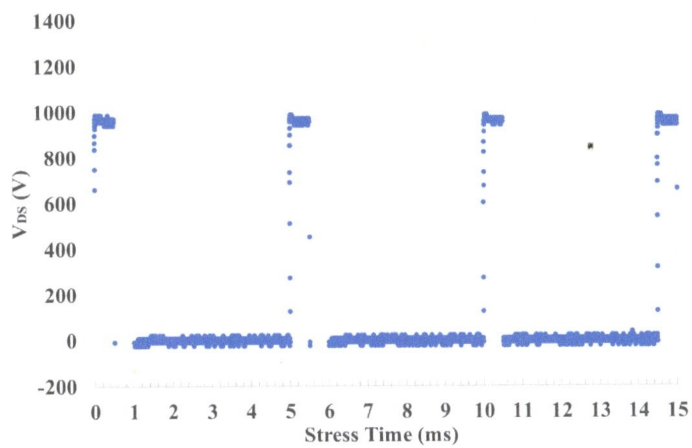

Figure 7. V_D stress pulse characterization.

4. Data/Results and Discussion

To simplify the graphical representation of drifting parameters under HTRB and HTGB stress, a limited selection of devices under testing has been reported in the figures related to the test. The experimental results have been confirmed on samples collected on different wafers belonging to different production lots. Some representative devices are reported in the following tables labeled as the Dev number belonging to their wafer (Wf) number.

4.1. HTRB Test

A preliminary characterization of the devices was performed before the HTRB test by taking into account the different parameters, shown in Figure 5 in orange boxes. Table 1 shows only the parameters, which, after stress, exhibit the most significant drift. It has to be emphasized that V_{th} and V_{FN-} (hole injections from p-body) are affected by the stress. The drift of the other electrical parameters is quite negligible.

Table 1. Channel- and gate oxide-related parameters before HTRB test and their corresponding drift (I(t)/I(0)) after 40 h of HTRB stress.

	V_{th} (V) @ I_{DS} = 5 µA V_{DS} = 0.1 V	V_{FN+} (V) @ I_{GS} = +10 nA	V_{FN-} (V) @ I_{GS} = −10 nA	BV_{DSS} (V) @ I_{DS} = 1 mA G & S Grounded	
DEV33–WF1	0.523	33.42	−17.30	879.11	Parametric data Before HTRB stress
DEV36-WF1	0.510	33.58	−17.78	881.43	
DEV19-WF2	0.814	36.00	−18.59	853.08	
DEV20-WF2	0.842	35.82	−18.14	853.34	
DEV33–WF1	0.989	1.01	1.07	1.00	Drift analysis After HTRB stress
DEV36-WF1	0.950	1.01	1.05	1.00	
DEV19-WF2	1.014	1.01	1.06	1.00	
DEV20-WF2	1.010	1.01	1.07	1.00	

4.2. HTGB Test

The same preliminary characterization was performed in HTGB stress conditions and the most significant results are, respectively, shown in Table 2 for the first read-out and for drift analysis.

Table 2. Channel- and gate oxide-related parameters before HTRB test and their corresponding drift (I(t)/I(0)) after 40 h HTGB stress.

	V_{th} (V) @ I_{DS} = 5 µA V_{DS} = 0.1 V	V_{FN+} (V) @ I_{GS} = +10 nA	V_{FN-} (V) @ I_{GS} = −10 nA	BV_{DSS} (V) @ I_{DS} = 1 mA G & S Grounded	
DEV34–WF1	0.45	33.68	−17.59	879.82	Parametric data Before HTGB stress
DEV37-WF1	0.51	32.98	−17.38	878.16	
DEV21-WF2	0.89	36.21	−18.41	840.66	
DEV22-WF2	0.82	35.46	−18.55	844.32	
DEV34–WF1	0.96	1.01	1.16	1.01	Drift analysis After HTGB stress
DEV37-WF1	0.91	1.01	1.15	1.01	
DEV21-WF2	0.99	1.01	1.11	1.01	
DEV22-WF2	1.00	1.01	1.10	1.01	

At the end of the characterization protocol, a data comparison of the drift analysis of HTRB-stressed devices and negative equivalent bias under HTGB stress was carried out on V_{th} (Figure 8). The decreasing V_{th} on the blue sample (stressed under HTGB conditions) is higher than that on the red ones (stressed under HTRB conditions). This is usually related to negative charge emission or hole trapping in the gate oxide. It can be noticed that the HTRB caused a variation in the threshold voltage of ΔV_{th} < 4% from the initial value. On the other hand, the HTGB caused a large variation (with respect of the HTRB) in the threshold voltage of ΔV_{th} < 15% from the initial value.

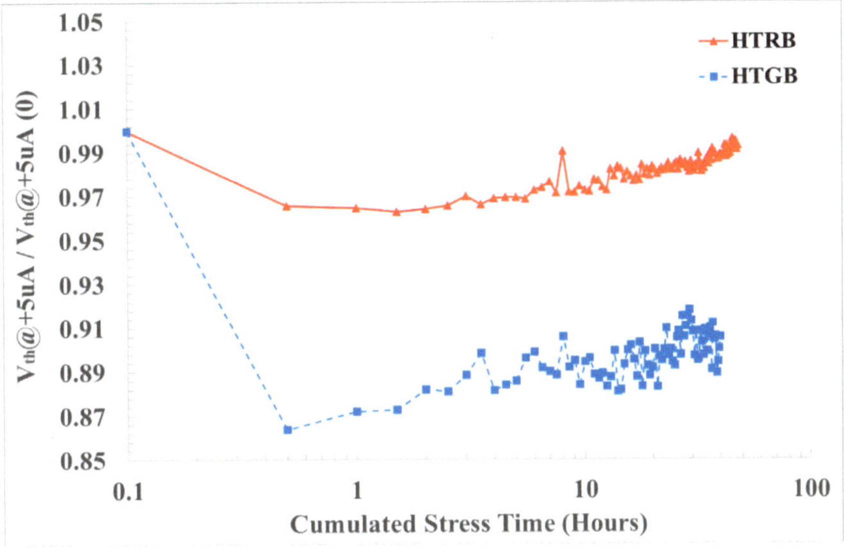

Figure 8. Threshold voltage drift under HTRB at 890 V (red) and HTGB at −19.5 V (blue) stress.

To confirm these results, and to emphasize the contribution of the charge trapped inside the gate oxide bulk, the V_{FN} drift under positive and negative bias has been put under analysis. The most significant drift, reported in Figure 9, is related to V_{FN-}, which is mainly affected by hole injection from the body and electrons from poly. Comparing the results with HTRB equivalent stress, the result is more evident than for V_{th} drift and the V_{FN}- shift is higher on the HTGB-stressed samples.

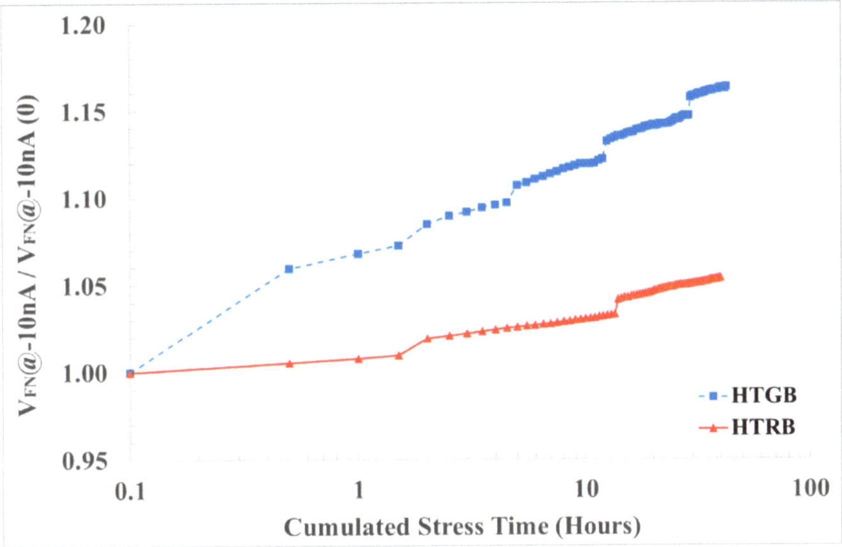

Figure 9. Negative F-N voltage onset drift under HTRB at 890 V (red) and HTGB at −19.5 V (blue) stress.

It can be noticed that the HTRB produced a variation in the onset voltage of the Fowler–Nordheim conduction of $\Delta V_{FN} < 4\%$ from the initial value. On the other hand, the

HTGB produced a large variation (with respect of the HTRB) in the threshold voltage ΔV_{FN} of < 15% from the initial value.

This experimental evidence can been understood by looking at the TCAD simulation in Figure 4B,C. In fact, in HTGB (Figure 4B) in the source region, the electric field is larger than that in the field in the JFET region; meanwhile, in HTRB (Figure 4C) in the source region, the electric field is a small fraction of that in the JFET region. Hence, different device degradation can be attributed to the different charge trapping occurring in the gate oxide in the proximity of the source region.

Finally, we calculated and quantified, by the charge-sensing method [30], the averaged trapped charge density inside the gate oxide. This confirmed the hypothesis of the worst-case effect on gate oxide degradation being that of negative bias gate stress as opposed to HTRB stress, even if the simulated gate field is the same (Figure 10). Negative bias stress causes a negative shift in V_{th}, which might be explained by the filling and emptying of near-interfacial oxide traps in the presence of a gate bias stress [31].

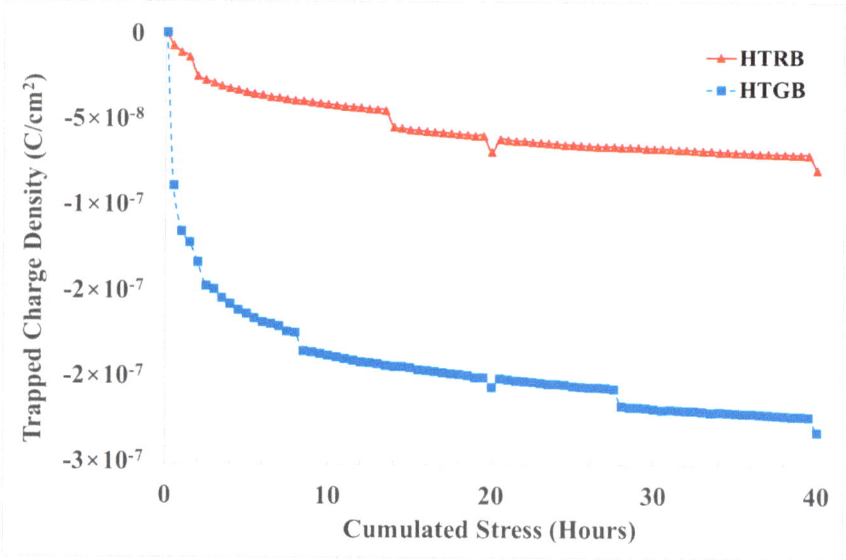

Figure 10. Oxide trapped charge density under HTRB (red) and HTGB (blue) stress.

5. Conclusions

In the field of device reliability, it is fundamental to evaluate which is the experimental procedure that represents the best trade-off between real-world application conditions and an accelerated characterization test to qualify material processing for device fabrication. Therefore, in this paper, a comparison of pulsed HTGB and HTRB stress has been carried out on defect-free 4H-SiC PowerMOSFETs at EWS in order to identify the most sensitive electrical test parameter under different stress conditions. From a wide number of analyzed electrical parameters, V_{th} and V_{FN-} showed higher instability and degradation compared to the others. Moreover, it has been demonstrated from the cross-correlation, the TCAD simulation and the experimental ΔV_{th} and ΔV_{FN-} variation that HTGB stress is more severe compared to HTRB stress. In fact, HTGB showed a 15% variation in both ΔV_{th} and ΔV_{FN-}; meanwhile, HTRB showed only a 4% variation in both ΔV_{th} and ΔV_{FN-} after an accumulated stress of 40 h. The physical explanation was attributed to the accelerated degradation of the gate insulator in the proximity of the source region under HTGB configuration.

Author Contributions: Conceptualization, L.A.; Methodology, W.C. and P.F.; Software, E.Z.; Investigation, L.A., E.Z., W.C. and P.F.; Data curation, P.F.; Writing—original draft, L.A.; Writing—review

& editing, P.F.; Visualization, F.R.; Supervision, A.R. and F.R.; Project administration, F.R.; Funding acquisition, A.R. and P.F. All authors have read and agreed to the published version of the manuscript.

Funding: The AdvanSiC project has received funding from the European Union's Horizon Europe programme under grant agreement no. 101075709.

Data Availability Statement: Data are contained within the article.

Acknowledgments: This paper has been partially supported by the Horizon EU Advances in Cost-Effective HV SiC Power Devices for Europe's Medium Voltage Grids (AdvanSiC). The views and opinions expressed are those of the author(s) only and do not necessarily reflect those of the European Union. Neither the European Union nor the granting authority can be held responsible for them.

Conflicts of Interest: Authors Laura Anoldo, Edoardo Zanetti, Walter Coco, Alfio Russo are employed by STMicroelectronics. The remaining authors declare that the research was conducted in the absence of any commercial or financial relationships that could be construed as a potential conflict of interest.

References

1. Kimoto, T.; Cooper, J.A. *Fundamentals of Silicon Carbide Technology: Growth, Characterization, Devices and Applications*; John Wiley Sons Singapore Pre. Ltd.: Singapore, 2014.
2. Gil, B.; Keshavarz, M.; Wales, D.; Darzi, A.; Yeatman, E. Orthogonal Surface-Enhanced RamanScattering/Field-Effect Transistor Detectionof Breast and Colorectal Cancer-Derived Exosomesusing Graphene as a Tag-Free Diagnostic Template. *Adv. NanoBiomed Res.* **2023**, *3*, 2300055. [CrossRef]
3. Kimoto, T. Material science and device physics in SiC technology for high-voltage power devices. *Jpn. J. Appl. Phys.* **2015**, *54*, 040103. [CrossRef]
4. Langpoklakpam, C.; Liu, A.-C.; Chu, K.-H.; Hsu, L.-H.; Lee, W.-C.; Chen, S.-C.; Sun, C.-W.; Shih, M.-H.; Lee, K.-Y.; Kuo, H.-C. Review of Silicon Carbide Processing for Power MOSFET. *Crystals* **2022**, *12*, 245. [CrossRef]
5. Guo, X.; Xun, Q.; Li, Z.; Du, S.J.M. Silicon Carbide Converters and MEMS Devices for High-temperature Power Electronics: A Critical Review. *Micromachines* **2019**, *10*, 406. [CrossRef] [PubMed]
6. Lichtenwalner, D.J.; Sabri, S.; van Brunt, E.; Hull, B.; Ryu, S.H.; Steinmann, P.; Romero, A.; Ganguly, S.; Gajewski, D.A.; Allen, S.; et al. Accelerated Testing of SiC Power Devices. In Proceedings of the 2020 IEEE International Integrated Reliability Workshop (IIRW), South Lake Tahoe, CA, USA, 9–14 October 2022.
7. Yang, L.; Castellazzi, A. High temperature gate-bias and reverse-bias tests on SiC MOSFETs. *Microelectron. Reliab.* **2013**, *53*, 1771–1773. [CrossRef]
8. Yano, H.; Kanafuji, N.; Osawa, A.; Hatayama, T.; Fuyuki, T. Threshold voltage instability in 4H-SiC MOSFETs with phosphorus-doped and nitrided gate oxides. *IEEE Trans. Electron. Dev.* **2015**, *62*, 324. [CrossRef]
9. Lelis, A.J.; Habersat, D.; Green, R.; Ogunniyi, A.; Gurfinkel, M.; Suehle, J.; Goldsman, N. Time dependence of bias-stress-induced SiC MOSFET threshold-voltage instability measurements. *IEEE Trans. Electron. Dev.* **2008**, *55*, 1835. [CrossRef]
10. Lelis, A.J.; Green, R.; Habersat, D.B.; El, M. Basic mechanisms of threshold-voltage instability and implications for reliability testing of SiC MOSFETs. *IEEE Trans. Electron. Dev.* **2015**, *62*, 316. [CrossRef]
11. Cabello, M.; Soler, V.; Rius, G.; Montserrat, J.; Rebollo, J.P. Godignon; Advanced processing for mobility improvement in 4H-SiC MOSFETs: A review. *Mater. Sci. Semicond. Process.* **2018**, *78*, 22–31. [CrossRef]
12. Singh, R. Reliability and performance limitations in SiC power devices. *Microelectron. Reliab.* **2006**, *46*, 713–730. [CrossRef]
13. Van Brunt, E.; Lichtenwalner, D.J.; Park, J.H.; Ganguly, S.J.W. McPherson; Lifetime Modeling of the 4H-SiC MOS Interface in the HTRB Condition Under the Influence of Screw Dislocations. In Proceedings of the 2023 IEEE International Reliability Physics Symposium (IRPS), Monterey, CA, USA, 26–30 March 2023.
14. Ganguly, S.; Lichtenwalner, D.J.; Isaacson, C.; Gajewski, D.A.; Steinmann, P.; Foarde, R.; Hull, B.; Ryu, S.-H.; Allen, S.; Palmour, J.W. Negative Gate Bias TDDB evaluation of n-Channel SiC Vertical Power MOSFETs. In Proceedings of the 2022 IEEE International Reliability Physics Symposium (IRPS) 8B. 1-1-8B, Dallas, TX, USA, 27–31 March 2022; pp. 1–6.
15. Schroder, D.K. *Semiconductor Material and Device Characterization*, 3rd ed.; Wiley: Hoboken, NJ, USA, 2006.
16. Aichinger, T.; Rescher, G.; Pobegen, G. Threshold voltage peculiarities and bias temperature instabilities of SiC MOSFETs. *Microelectron. Reliab.* **2018**, *80*, 68–78. [CrossRef]
17. Moens, P.; Geenen, F.; De Schepper, L.; Cano, J.F.; Lettens, J.; Maslougkas, S.; Franchi, J.; Domeij, M. The Concept of Safe Operating Area for Gate Dielectrics: The SiC/SiO_2 Case Study. In *IEEE International Reliability Physics Symposium Proceedings*; IEEE: Piscataway, NJ, USA, 2023.
18. Masin, F.; De Santi, C.; Lettens, J.; Franchi, J.; Domeij, M.; Moens, P.; Zanoni, E. Non-monotonic threshold voltage variation in 4H-SiC MOSFET: Investigation and modeling. *JAP* **2021**, *130*, 7285. [CrossRef]
19. Zhu, S.; Liu, T.; White, M.H.; Agarwal, A.K.; Salemi, A.; Sheridan, D. Investigation of gate leakage current behavior for commercial 1.2 kV 4H-SiC power mosfets. In Proceedings of the IRPS 2021, Monterey, CA, USA, 21–25 March 2021; pp. 1–7.

20. Green, R.; Lelis, A.; Habersat, D. Threshold-voltage biastemperature instability in commercially-available SiC MOSFETs. *Jpn. J. Appl. Phys.* **2016**, *55*, 4. [CrossRef]
21. Rescher, G.; Pobegen, G.; Grasser, T. Threshold voltage instabilities of present SiC-power MOSFETs under positive bias temperature stress. *Mater. Sci. Forum.* **2016**, *858*, 481–484. [CrossRef]
22. Fiorenza, P.; Zignale, M.; Camalleri, M.; Scalia, L.; Zanetti, E.; Saggio, M.; Giannazzo, F.; Roccaforte, F. Impact of the NO annealing duration on the SiO2/4H–SiC interface properties in lateral MOSFETs: The energetic profile of the near-interface-oxide traps. *Mater. Sci. Semicond. Process.* **2024**, *169*, 107866. [CrossRef]
23. El Hageali, S.A.; Guthrey, H.; Johnston, S.; Soto, J.; Odekirk, B.; Gorman, B.P. Mowafak Al-Jassim; Nondestructive microstructural investigation of defects in 4H-SiC epilayers using a multiscale luminescence analysis approach. *J. Appl. Phys.* **2022**, *131*, 185705. [CrossRef]
24. Fiorenza, P.; La Magna, A.; Vivona, M.; Roccaforte, F. Near interface traps in SiO_2/4H-SiC metal-oxide-semiconductor field effect transistors monitored by temperature dependent gate current transient measurements. *Appl. Phys. Lett.* **2016**, *109*, 012102. [CrossRef]
25. Fiorenza, P.; Vivona, M.; Iucolano, F.; Severino, A.; Lorenti, S.; Nicotra, G.; Bongiorno, C.; Giannazzo, F.; Roccaforte, F. Temperature-dependent Fowler-Nordheim electron barrier height in SiO_2/4H-SiC MOS capacitors. *Mater. Sci. Semicond. Process.* **2018**, *78*, 38–42. [CrossRef]
26. Chanana, R.K.; McDonald, K.; Di Ventra, M.; Pantelides, S.T.; FeldmanG, L.C.; Chung, Y.; Tin, C.C.; Williams, J.R.; Weller, R.A. Fowler–Nordheim hole tunneling in p-SiC/SiO_2 structures. *Appl. Phys. Lett.* **2000**, *77*, 2560–2562. [CrossRef]
27. Fiorenza, P.; Cordiano, F.; Alessandrino, S.M.; Russo, A.; Zanetti, E.; Saggio, M.; Bongiorno, C.; Giannazzo, F.; Roccaforte, F. Consideration on the extrapolation of the low insulator field TDDB in 4H-SiC power MOSFETs. In Proceedings of the 2023 IEEE International Reliability Physics Symposium (IRPS), Monterey, CA, USA, 26–30 March 2023; pp. 1–4. [CrossRef]
28. Liu, T.; Zhu, S.; Yu, S.; Xing, D.; Salemi, A.; Kang, M.; Booth, K.; White, M.H.; Agarwal, A.K. Gate oxide reliability studies of commercial 1.2 kV 4H-SiC power MOSFETs. In Proceedings of the IRPS 2020, Dallas, TX, USA, 28 April–30 May 2020; pp. 1–5.
29. Fayyaz, A.; Castellazzi, A. High temperature pulsed-gate robustness testing of SiC power MOSFETs. *Microelectron. Reliab.* **2015**, *55*, 1724–1728. [CrossRef]
30. DiMaria, D.J.; Stasiak, J.W. Trap creation insilicon dioxide produced by hot electrons. *J. Appl. Phys.* **1989**, *65*, 2342. [CrossRef]
31. Fiorenza, P.; Bongiorno, C.; Giannazzo, F.; Alessandrino, M.S.; Messina, A.; Saggio, M.; Roccaforte, F. Interfacial electrical and chemical properties of deposited SiO_2 layers in lateral implanted 4H-SiC MOSFETs subjected to different nitridations. *Appl. Surf. Sci.* **2021**, *557*, 149752. [CrossRef]

materials

Article

Modeling of Charge-to-Breakdown with an Electron Trapping Model for Analysis of Thermal Gate Oxide Failure Mechanism in SiC Power MOSFETs

Jiashu Qian [1,*], Limeng Shi [1], Michael Jin [1], Monikuntala Bhattacharya [1], Atsushi Shimbori [2], Hengyu Yu [1], Shiva Houshmand [1], Marvin H. White [1] and Anant K. Agarwal [1,*]

1 Department of Electrical & Computer Engineering, The Ohio State University, Columbus, OH 43210, USA; shi.1564@osu.edu (L.S.); jin.845@osu.edu (M.J.); bhattacharya.119@osu.edu (M.B.); yu.3868@osu.edu (H.Y.); houshmand.3@osu.edu (S.H.); white.1829@osu.edu (M.H.W.)
2 Ford Motor Co., Dearborn, MI 48126, USA; ashimbor@ford.com
* Correspondence: qian.539@osu.edu (J.Q.); agarwal.334@osu.edu (A.K.A.)

Abstract: The failure mechanism of thermal gate oxide in silicon carbide (SiC) power metal oxide semiconductor field effect transistors (MOSFETs), whether it is field-driven breakdown or charge-driven breakdown, has always been a controversial topic. Previous studies have demonstrated that the failure time of thermally grown silicon dioxide (SiO_2) on SiC stressed with a constant voltage is indicated as charge driven rather than field driven through the observation of Weibull Slope β. Considering the importance of the accurate failure mechanism for the thermal gate oxide lifetime prediction model of time-dependent dielectric breakdown (TDDB), charge-driven breakdown needs to be further fundamentally justified. In this work, the charge-to-breakdown (Q_{BD}) of the thermal gate oxide in a type of commercial planar SiC power MOSFETs, under the constant current stress (CCS), constant voltage stress (CVS), and pulsed voltage stress (PVS) are extracted, respectively. A mathematical electron trapping model in thermal SiO_2 grown on single crystal silicon (Si) under CCS, which was proposed by M. Liang et al., is proven to work equally well with thermal SiO_2 grown on SiC and used to deduce the Q_{BD} model of the device under test (DUT). Compared with the Q_{BD} obtained under the three stress conditions, the charge-driven breakdown mechanism is validated in the thermal gate oxide of SiC power MOSFETs.

Keywords: thermal gate oxide; SiC; MOSFET; charge-driven breakdown; Q_{BD}; CCS; CVS; PVS; electron trapping model

check for updates

Citation: Qian, J.; Shi, L.; Jin, M.; Bhattacharya, M.; Shimbori, A.; Yu, H.; Houshmand, S.; White, M.H.; Agarwal, A.K. Modeling of Charge-to-Breakdown with an Electron Trapping Model for Analysis of Thermal Gate Oxide Failure Mechanism in SiC Power MOSFETs. *Materials* **2024**, *17*, 1455. https://doi.org/10.3390/ma17071455

Academic Editors: Marilena Vivona and Mike Jennings

Received: 5 February 2024
Revised: 12 March 2024
Accepted: 20 March 2024
Published: 22 March 2024

1. Introduction

Silicon carbide (SiC) power MOSFETs are gradually gaining market attention due to their lower switching losses, higher temperature capability, higher switching frequencies, and increasingly competitive price compared to their silicon (Si) counterparts [1,2]. Especially in the field of electric vehicles (EVs), the aforementioned advantages make them largely attractive to EV OEMs and tier-one suppliers for potential applications in onboard chargers and drivetrain inverters [3–5]. Planar SiC power MOSFETs, with their relatively more mature process and cheaper manufacturing costs, have become the mainstream commercial SiC power MOSFETs on the market [6–11]. Trench SiC power MOSFETs, although optimized in device performance due to enhanced electron mobility and elimination of JFET resistance, as well as smaller cell pitch, still hold a relatively small market share due to their higher cost and lower reliability [12,13]. The lower reliability is mainly caused by electric field crowding at the corner of the trench gate and implantation-induced basal plane dislocation (BPD) [14–17]. Therefore, although the performance and structural limitations of planar SiC power MOSFETs are gradually becoming apparent, unless trench SiC power MOSFETs with better economy and reliability are commercialized, the main way to improve the performance of planar SiC power MOSFETs is to operate the devices at a higher gate oxide field to increase the channel electron density [18,19]. However, this places more stringent demands on the reliability of gate oxide in planar SiC power MOSFETs. One major area of concern is that the prediction of gate oxide lifetime under the typical operation, with an increased gate oxide field, still needs to meet conservative design requirements [20]. This requires not only improvements to the thermal growth process of gate oxide in planar SiC power MOSFETs, but also sufficient accuracy in the gate oxide lifetime prediction model [21]. The key to determining the accuracy of the prediction model is the failure mechanism of thermal gate oxide grown on SiC [22].

The commonly used gate oxide lifetime prediction method in the industry for planar SiC power MOSFETs is the time-dependent dielectric breakdown (TDDB) test. As MOSFETs are voltage-controlled devices, the conventional TDDB test in the industry is constant voltage stress TDDB (CVS-TDDB) based on the thermochemical E model, as it provides the most conservative lifetime extrapolation, even without physical or even experimental justification [23–25]. The thermochemical E model is considered to be based on the thermal gate oxide failure mechanism of field-driven breakdown [26]. P. Moens et al. questioned this mechanism and proposed a more rational failure mechanism of charge-driven breakdown [27]. The team grew approximately 53 nm of silicon dioxide (SiO_2) on n-epi SiC to form circular capacitor structures with n+ doped polysilicon gates. By measuring the gate leakage current as a function of the gate oxide field at different temperatures, they concluded that the electron tunneling mechanism from SiC to SiO_2 transitions from thermally assisted tunneling (TAT) to Fowler–Nordheim tunneling (FNT) as the gate oxide field increases. During the transition, there is a phase where both the electron tunneling mechanisms jointly influence. This transition in the electron tunneling mechanism cannot be described in the conventional Weibull plots of CVS-TDDB based on field-driven breakdown but can be accurately depicted in new Weibull plots based on charge-driven breakdown, where the Weibull Slope (β) at the TAT dominant stage, FNT dominant stage, and the joint influence stage each have a specific value that decreases in the stage order of TAT dominance, joint influence, and FNT dominance. Therefore, the failure mechanism of thermal gate oxide grown on SiC is considered to be charge-driven breakdown rather than field-driven breakdown, and a more optimistic lifetime prediction model based on Q_{BD} has been proposed. Since the stressor is charge rather than field, constant current stress (CCS) is considered as a better stress method because it is not negatively affected by trapped electrons in the gate oxide and can reach Q_{BD} faster [28]. The β value of the Weibull plots based on the CCS-Q_{BD} approach has also been proven to accurately describe the transition in the electron tunneling mechanism.

This work draws on the electron trapping model in very thin SiO_2 (no more than 10 nm) thermally grown on Si under CCS by M. Liang et al., proving its applicability also in approximately 4–5 times thicker SiO_2 thermally grown on SiC through CCS-TDDB tests on thermal gate oxide in a type of commercial planar SiC power MOSFETs until failure [29]. Based on this electron trapping model, a Q_{BD} model for thermal gate oxide in the commercial planar SiC power MOSFETs under test is established. Considering that the gate voltage signal for MOSFETs is a pulse-width-modulated (PWM) signal rather than a constant in actual applications, MOSFETs are voltage-controlled devices [30]. Therefore, in addition to conventional CVS, this paper also extracts the Q_{BD} of the thermal gate oxide in commercial planar SiC power MOSFETs under pulsed voltage stress (PVS) and CCS at different stress levels for comparison with the established Q_{BD} model. The high match between the extracted Q_{BD} and the Q_{BD} model indicates that different stress methods do not change the failure mechanism of thermal gate oxide, and the existence of a specific Q_{BD} that causes the thermal gate oxide to fail under different stress methods further proves that charge-driven breakdown is the failure mechanism of thermal gate oxide. Additionally, the lifetime prediction model established based on this failure mechanism can be considered more credible even if it is not as conservative as the thermochemical E model [31]. This will also provide a theoretical basis for suggesting the industry adopt more aggressive screening methods to more effectively screen out extrinsic defects in thermal gate oxide according to the more optimistic lifetime prediction [32].

2. Materials and Methods

2.1. Devices under Test (DUTs)

In this work, the devices under test (DUTs) are commercial 1200 V planar SiC power MOSFETs packaged in TO-247-3 from Vendor E. Considering that as more and more EV OEMs upgrade from 400 V systems to 800–900 V systems, the voltage rating of 1200 V will receive more attention from the market, so selecting this DUT is a better reference for the industry [33–37]. The curves of gate leakage current (I_{gss}) for three DUTs at 150 °C, as a function of gate voltage (V_g), are presented in Figure 1. The high overlap of the three curves demonstrates the high uniformity in gate oxide quality of the commercial DUTs. This indicates that these commercial DUTs undergo stringent gate oxide screening before leaving the factory, reducing the adverse impact of early oxide failure caused by extrinsic defects on subsequent test results [38]. The I_{gss} curves for all three DUTs exhibit breakdown near 50 V, with an average gate oxide breakdown voltage of about 48.57 V. Based on the assumption that the critical breakdown electric field is about 11 MV/cm for SiO_2, the gate oxide thickness of DUTs can be estimated to be approximately 44.15 nm [39]. According to the total capacitance (C_{tot}) derived from the gate C-V measurements of DUTs and C_{ox} of SiO_2, the gate oxide area in each DUT is estimated to be about 0.9 mm^2. General information of the commercial DUTs used in this work is summarized in Table 1.

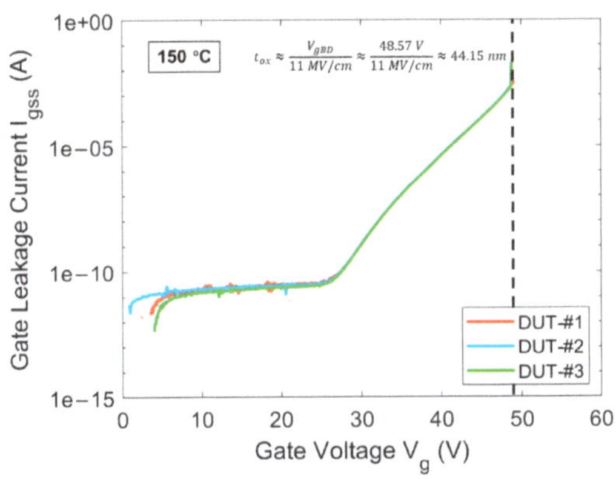

Figure 1. I_{gss} curves as a function of V_g at 150 °C until oxide breakdown for three DUTs. The dashed line indicates the oxide breakdown voltage.

Table 1. General information of DUTs in this work.

Vendor	Voltage Rating (V)	Current Rating (A)	Structure	Est. Oxide Thickness (nm)	Est. Oxide Area (mm^2)
E	1200	11	Planar	44.15	0.9

2.2. Experimental Methods

2.2.1. Liang and Hu's Electron Trapping Model

A mathematical model for describing the electron trapping phenomenon in very thin SiO_2 thermally grown on Si under CCS has been proposed by M. Liang and C. Hu [29]. In this model, M. Liang et al. have demonstrated that when the thickness of SiO_2 in a polycrystalline-Si-SiO_2-Si MOS capacitor structure reaches a certain level, the change in V_g (ΔV_g) between the polycrystalline-Si gate and the grounded Si measured under CCS tends to saturate at a high electron fluence (F). However, in the case of thinner SiO_2, ΔV_g does not show a saturation trend with F but instead tends to linearly increase until the oxide breakdown. This phenomenon is also observed under various CCS, and with different thicknesses of SiO_2, as long as they do not exceed the critical oxide thickness. Therefore, for thinner SiO_2, M. Liang et al. believe that in addition to the pre-existing electron traps in SiO_2, new electron traps are being generated during CCS. The pre-existing electron traps and the generated electron traps, having different trap capture cross-sections and trap centroids, collectively capture electrons tunneling from Si into the oxide, thus affecting ΔV_g. Based on this, a comprehensive mathematical model is established and used to characterize the electron trapping phenomenon in the 100 Å SiO_2 of a fabricated Si MOS capacitor structure.

In this model, the density of filled electron traps can be expressed as follows:

$$N_{ot}(\sigma_p,\sigma_g,F) = N_{opt}(\sigma_p,F) + N_{ogt}(\sigma_g,F) = N_{op}(\sigma_p)\left(1-e^{-\sigma_p F}\right) + q\frac{g}{J}[F - \frac{1}{\sigma_g}(1-e^{-\sigma_g F})] \tag{1}$$

where

N_{ot}—density of filled electron traps;
N_{op}/N_{opt}—density of pre-existing total/filled electron traps;
N_{ogt}—density of filled generated electron traps;
σ_p/σ_g—capture cross-section of pre-existing/generated electron traps;
q—electric charge of an electron;
J—current density of the specific CCS;
F—electron fluence ($F = J \cdot t/q$, t is the stress time under the specific CCS);
g—generation rate of generated electron traps under the specific CCS.

Therefore, ΔV_g due to the filled electron traps can be expressed as follows:

$$\Delta V_g(F) = \frac{q}{\varepsilon_{ox}}\overline{x}(F) N_{ot}(\sigma_p,\sigma_g,F) \tag{2}$$

where ε_{ox} is the dielectric constant of SiO_2 and $\overline{x}$ is the centroid of electron traps measured from the gate. Figure 2 presents a method for extracting $\overline{x}$ with respect to F through shifts in I_g-V_g curves at different stages under a specific CCS as shown below.

Also, $\overline{x}$ can be represented by the centroid of pre-existing electron traps ($\overline{x_p}$) and the centroid of generated electron traps ($\overline{x_g}$) as follows:

$$\overline{x}(F) = \frac{\overline{x_p}N_{opt} + \overline{x_g}N_{ogt}}{N_{opt} + N_{ogt}} \tag{3}$$

When F is large enough under a specific CCS, ΔV_g can be simplified to the following:

$$\Delta V_g(F) = \frac{q}{\varepsilon_{ox}}\left[\overline{x_p}N_{op} - \overline{x_g}q\frac{g}{J}\frac{1}{\sigma_g} + \overline{x_g}q\frac{g}{J}F\right] = \frac{q}{\varepsilon_{ox}}\overline{x_g}q\frac{g}{J}\cdot F + \frac{q}{\varepsilon_{ox}}\left(\overline{x_p}N_{op} - \overline{x_g}q\frac{g}{J}\frac{1}{\sigma_g}\right) \tag{4}$$

Considering that $\overline{x_p}N_{op}$ is a constant characteristic value regarding pre-existing electron traps and the generation rate of generated electron traps g under a specific CCS is also considered as a specific constant value in the model, Equation (4) can be regarded as a linear expression of ΔV_g with respect to F when F is large enough. Moreover, differentiating Equation (4) can give the constant slope of this linear expression as follows:

$$\frac{d\Delta V_g}{dF} = \frac{q}{\varepsilon_{ox}}\overline{x_g}q\frac{g}{J} \tag{5}$$

From Equation (3), it is known that $\overline{x}$ varies due to the ratio change between N_{opt} and N_{ogt} under different F. When F is large enough, N_{opt}, having tended to saturate earlier, becomes almost negligible relative to N_{ogt}, which continues to increase with the constant generation rate of new electron traps. In this case, $\overline{x}$ tends to saturate, and the saturation value approached can be estimated as $\overline{x_g}$. In the model, $\overline{x_g}$ is found to be a constant value, unaffected by CCS. This phenomenon is also reflected in the measurements of gate oxide in commercial SiC DUTs in this work.

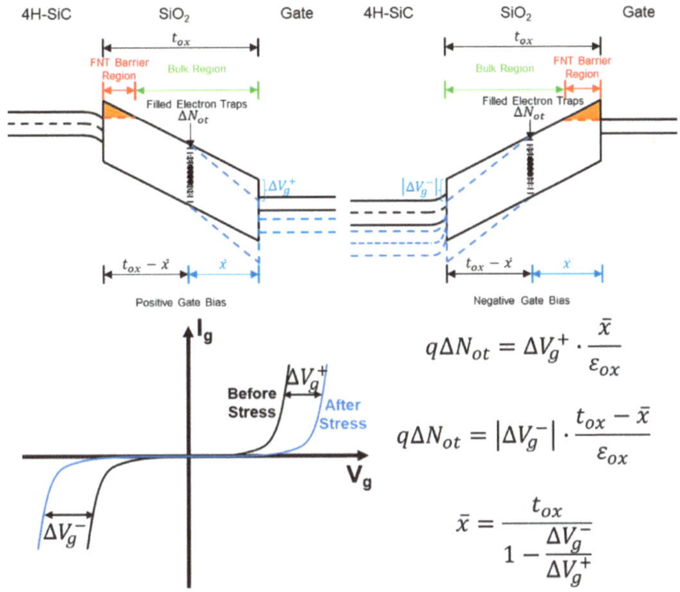

Figure 2. Energy band variation caused by ΔV_g to maintain a constant FNT barrier for a constant $|I_{gss}|$. $\bar{x}$ can be extracted based on ΔV_g at different stages under the constant $|I_{gss}|$.

2.2.2. Extraction of Charge-to-Breakdown (Q_{BD})

Q_{BD} measurement is a standard destructive method used to determine the quality of gate oxide in MOS devices. Q_{BD} is extracted by calculating the total charge passing through the dielectric (i.e., the product of total electron fluence and the electric charge of an electron, or the integral of electron current over time-to-breakdown (t_{BD}), making it a time-dependent measurement [27]. The extraction of Q_{BD} can be represented as follows:

$$Q_{BD} = q \cdot F \cdot A_{ox} = \int_0^{t_{BD}} I(t)d(t) \tag{6}$$

3. Results

3.1. Modeling of ΔV_g When Breakdown Occurs (ΔV_{gBD}) in Commercial SiC DUTs

3.1.1. $\overline{x_g}$ Extraction

In Figure 3, based on the above method of extracting $\bar{x}$, the curves of $\bar{x}$ versus F for the commercial SiC DUTs at 150 °C under a CCS of 0.5 and 0.7 μA are shown. It is observable that the two curves highly coincide, consistent with what is measured in the oxide thermally grown on Si that there is no correlation with the CCS. However, due to the inferior quality of oxide thermally grown on SiC compared to Si, the oxide fails before F is large enough for $\bar{x}$ to reach its saturation value [40]. Therefore, by fitting the overlapped curves of $\bar{x}$ versus F, the $\overline{x_g}$ of DUTs is estimated to be approximately 16.5 nm measured from the gate.

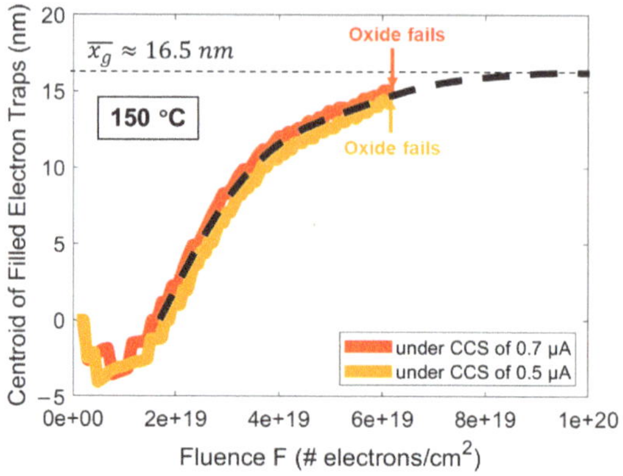

Figure 3. $\overline{x}$ curves as a function of F for DUTs at 150 °C under a CCS of 0.5 and 0.7 μA, respectively, using the extraction method introduced in Figure 2.

3.1.2. Mathematical Expression of ΔV_{gBD}

In Figure 4, the curves of V_g over stress time until the oxide breakdown at 150 °C for six DUTs under a CCS of 0.7 μA are shown. ΔV_g can be obtained by subtracting the initial V_g from V_g at different time points. Multiplying time by the known current density under CCS and dividing by the electric charge of an electron yields the electron fluence. Figure 5a presents the curves of ΔV_g versus F until the oxide breakdown at 150 °C for the six DUTs under a CCS of 0.7 μA. Differentiating the curves in Figure 5a results in the curves shown in Figure 5b. The high consistency among the six curves in both again proves the uniformity of the oxide quality in these commercial SiC DUTs after a possible stringent gate oxide screening. According to Equation (4), the electron trapping phenomenon in the oxide of these commercial SiC DUTs shows characteristics similar to those predicted by the model for very thin oxide thermally grown on Si. By extending the linear part of the curves within the high F range in Figure 5a to intersect with the y-axis, the value of intersection point is estimated to be approximately −0.7 V. Additionally, the saturation value extracted in Figure 5b within the corresponding F range for the linear part of the curves in Figure 5a is about 6.76×10^{-20} V·cm^2. Therefore, the relevant mathematical expressions can be represented as follows:

$$\frac{q}{\varepsilon_{ox}}\left(\overline{x_p}N_{op} - \overline{x_g}q\frac{g}{J}\frac{1}{\sigma_g}\right) \approx -0.7\ \text{V} \tag{7}$$

$$\frac{q}{\varepsilon_{ox}}\overline{x_g}q\frac{g}{J} \approx 6.76 \times 10^{-20}\ \text{V}\cdot\text{cm}^2 \tag{8}$$

Since $J \approx 0.7\ \mu\text{A}/0.9\ \text{mm}^2 \approx 7.8 \times 10^{-5}\ \text{A}/\text{cm}^2$ and $\overline{x_g} \approx 16.5$ nm, the above expressions can be transformed into the following:

$$g(0.7\ \mu\text{A}) \approx 4.3 \times 10^7\ \text{cm}^{-2}\cdot\text{s}^{-1} \tag{9}$$

$$\overline{x_p}N_{op} - \frac{1.46 \times 10^{-13}}{\sigma_g} \approx -1.51 \times 10^6 \text{cm}^{-1} \tag{10}$$

Similarly, Figure 6a displays the curves of V_g over stress time until the oxide breakdown for three DUTs under CCS of 0.14 μA. Moreover, both the characteristics of ΔV_g versus F for three DUTs under a CCS of 0.14 μA shown in Figure 6b, and of the differentiated curves in Figure 6c, are very similar to those in Figure 5a,b. Therefore, by repeating the aforementioned method, similar relevant mathematical expressions can be obtained as

$$g(0.14\ \mu\text{A}) \approx 8.47 \times 10^6\ \text{cm}^{-2}\cdot\text{s}^{-1} \tag{11}$$

$$\overline{x_p}N_{op} - \frac{1.43 \times 10^{-13}}{\sigma_g} \approx -2.157 \times 10^5\text{cm}^{-1} \tag{12}$$

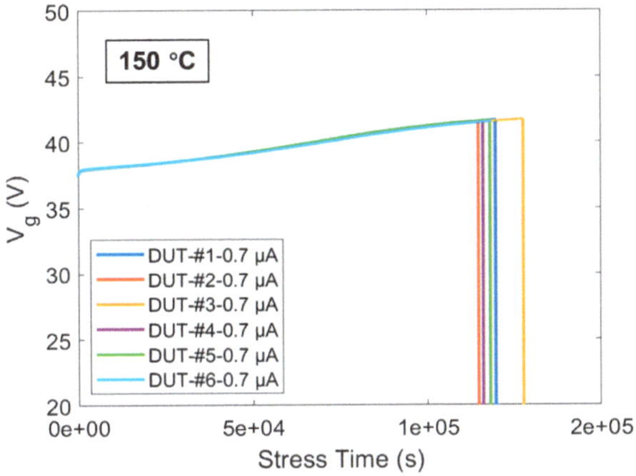

Figure 4. V_g curves as a function of stress time until oxide breakdown at 150 °C for six DUTs under a CCS of 0.7 μA.

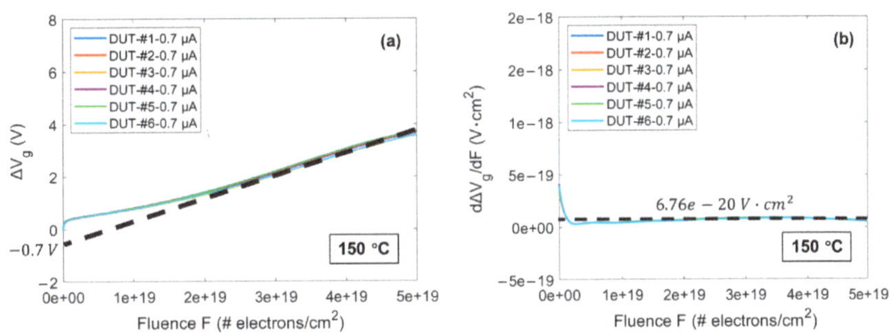

Figure 5. (**a**) ΔV_g curves as a function of F until oxide breakdown at 150 °C for the six DUTs under a CCS of 0.7 μA; (**b**) Differentiated curves from (**a**). # is a number sign representing the number of electrons.

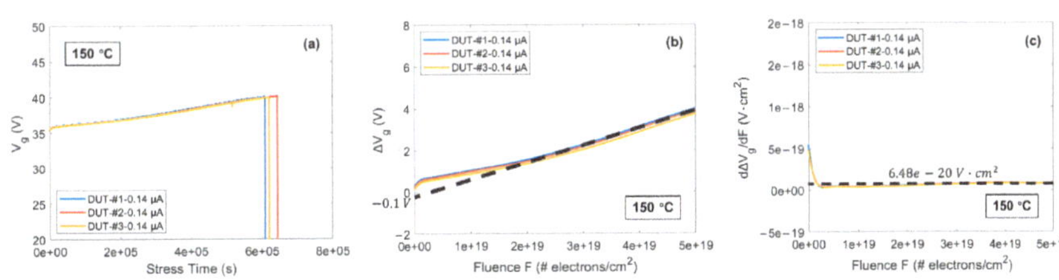

Figure 6. (**a**) V_g curves as a function of stress time until the oxide breakdown at 150 °C for three DUTs under CCS of 0.14 μA; (**b**) ΔV_g curves as a function of F until the oxide breakdown at 150 °C for the three DUTs under CCS of 0.14 μA; (**c**) Differentiated curves from (**b**).

Considering that the DUTs stressed under a CCS of 0.7 and 0.14 μA are from the same batch of identical devices produced on the same wafer using exactly the same process, $\overline{x_p}N_{op}$ can be considered a constant value unaffected by CCS. Also, in the model, the generated electron traps under CCS have been proven to have a centroid always at a specific and constant position unaffected by CCS, with CCS mainly affecting their generation rate. Furthermore, σ_g, as a specific attribute of the generated electron traps, is also considered to be a constant value unaffected by CCS. Therefore, relating Equations (10) and (12) can give $\overline{x_p}N_{op}$ and σ_g values of approximately 6.2×10^7 cm^{-1} and 2.317×10^{-21} cm^2. Since the electron-fluence-to-breakdown (F_{BD}) with respect to t_{BD} of the oxide can be expressed as $F_{BD} = J{\cdot}t_{BD}/q$, and with the CCS value $I \approx 0.009{\cdot}J$, the mathematical relationship between ΔV_{gBD} and I can be expressed as follows:

$$\Delta V_{gBD}(I) \approx 7.7 \times 10^{-13} \cdot g(I){\cdot}t_{BD} - 4.68 \times 10^{-13} \cdot \frac{g(I)}{I} + 28.74 \text{ V} \quad (13)$$

Figure 7 shows the curves of V_g over stress time until oxide breakdown for DUTs under all CCS scenarios used in this work. The applied CCS values include 23.2 nA, 0.14 μA, 0.275 μA, 0.7 μA, 3.43 μA, 15.94 μA, 19.5 μA, 34.3 μA, and 61.1 μA, corresponding to gate oxide electric fields of 7.5, 8, 8.2, 8.5, 9, 9.5, 9.6, 9.8, and 10 MV/cm, respectively, estimated by correlative V_g of CCS in Figure 1 divided by the oxide thickness. The t_{BD} of gate oxide in DUTs under each CCS can be extracted when the curves of V_g sharply drop and the average t_{BD} at 150 °C under each CCS are reflected in Figure 8. It can be observed that under CCS, t_{BD} follows a $1/I$ model, which can be expressed as follows:

$$t_{BD}(I) = A{\cdot}I^{-B} \quad (14)$$

where A and B are constant. For DUTs in this work, under CCS, $t_{BD}(I)$ is fitted by the $1/I$ model as follows:

$$t_{BD}(I) = 0.071 \cdot I^{-1.017}{\cdot}\text{s} \quad (15)$$

Or in the log-log scale, Equation (14) can be transformed into the following:

$$\log(t_{BD}(I)) \approx -1.017 \cdot \log(I) - 1.1492 \quad (16)$$

Which is in a linear relationship as shown in the inset of Figure 8.

Using the method described earlier for extracting g of generated electron traps under a specific CCS, g under each CCS is extracted and is presented in Figure 9. It can be observed that g follows a linear I model. The mathematical expression for this linear I model is as follows:

$$g(I) \approx 6.13 \times 10^{13} \cdot I - 30324.35 \text{ cm}^{-2} \cdot \text{s}^{-1} \quad (17)$$

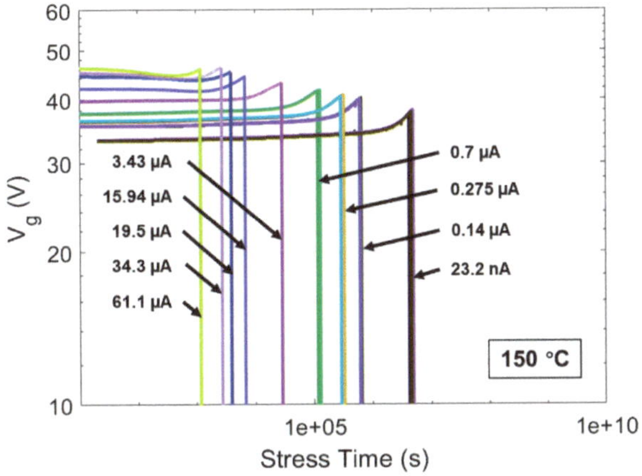

Figure 7. V_g curves as a function of stress time until oxide breakdown at 150 °C for multiple DUTs under CCS values of 23.2 nA, 0.14 µA, 0.275 µA, 0.7 µA, 3.43 µA, 15.94 µA, 19.5 µA, 34.3 µA, and 61.1 µA, respectively.

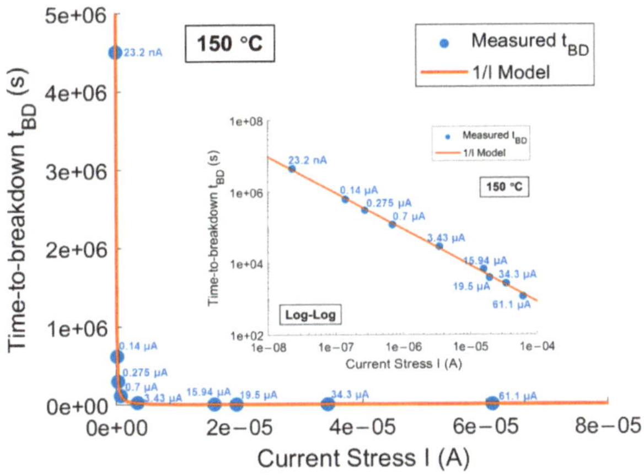

Figure 8. Average t_{BD} of gate oxide at 150 °C under each CCS for multiple DUTs fitted by a $1/I$ model. The inset shows the log–log scale with a linear relationship.

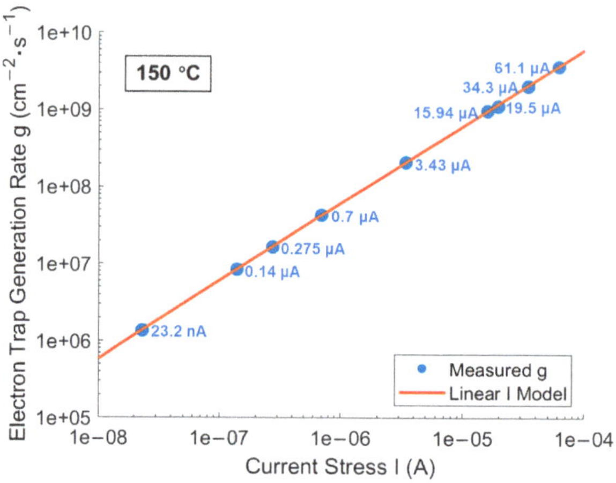

Figure 9. g extracted from the measured data at 150 °C under each CCS fitted by a linear I model.

Therefore, the mathematical expression of ΔV_{gBD} as a function of I can be summarized as the combination of Equations (13), (14), and (17). The curve of the mathematical expression is displayed in Figure 10 as model-based ΔV_{gBD}. Additionally, ΔV_{gBD} for DUTs under each CCS can be obtained from Figure 7 by subtracting the initial V_g from V_g at the point of gate oxide breakdown, which is also reflected in Figure 10. It is observed that the measured ΔV_{gBD} under all CCS values not exceeding 3.43 μA highly coincides with the curve of model-based ΔV_{gBD} as a function of I. However, as CCS gradually exceeds 3.43 μA, the measured ΔV_{gBD} starts to fall below the model expectation. This discrepancy arises because, under CCS not exceeding 3.43 μA, the electron tunneling mechanism is predominantly thermally assisted tunneling (TAT), with the oxide's trapped charge mainly consisting of electrons, making the electron trapping model applicable in this range. The tunneling electrons lack sufficient energy to trigger enough impact ionization, thus preventing trapped holes induced by anode hole injection (AHI) from dominating over trapped electrons. In contrast, when CCS exceeds 3.43 μA, the electron tunneling mechanism shifts more toward Fowler–Nordheim tunneling (FNT). In this regime, the tunneling electrons possess enough energy at the beginning to cause significant impact ionization, leading to a dominance of trapped holes in the oxide during the first stage of CCS, although trapped electrons subsequently regain dominance. Since the electron trapping model does not account for trapped holes and is solely based on trapped electrons, it is not applicable in the CCS range where trapped holes also play a role. This explanation is corroborated by the trends observed in Figure 7, where under CCS values up to 3.43 μA, the V_g curves consistently show an increasing trend due to electron trapping in the oxide throughout the entire stress to breakdown. In contrast, under CCS values exceeding 3.43 μA, the V_g curves initially show a decreasing trend due to hole trapping in the oxide, followed by a dominance of electron trapping leading to an increasing trend up to breakdown, and the initial decrease in the V_g curves becomes more pronounced as CCS increases beyond 3.43 μA. In summary, it can be concluded that Liang and Hu's electron trapping model, established for very thin (no more than 10 nm) thermally grown SiO_2 on Si, is equally applicable to thicker (up to 45 nm in this work) SiO_2 thermally grown on SiC. This finding will aid in developing a Q_{BD} model for the commercial SiC DUTs.

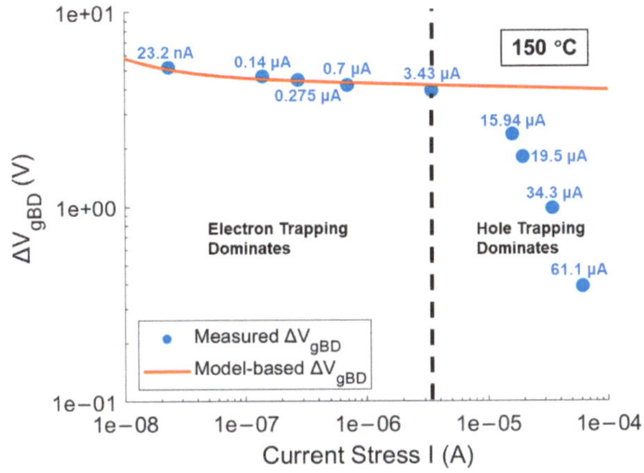

Figure 10. Comparison of model-based ΔV_{gBD} with measured ΔV_{gBD} at 150 °C.

3.2. Modeling of Q_{BD} in Commercial SiC DUTs

The V_g curves measured in Figure 11 show that the oxide breakdown points of the V_g curves under all CCS values follow a linear t_{BD} model on a log–log scale. The mathematical expression for this linear t_{BD} model can be represented as follows:

$$\log V_{gBD} = -0.0242 \cdot \log t_{BD} + 1.7448 \quad (18)$$

If the segment of the I_{gss} curves for DUTs in Figure 1, ranging from approximately 21 nA to 1.2 mA, is extracted as the current stress operating region, the corresponding V_g range is approximately 33 to 48 V. By adding ΔV_{gBD}, extracted using its mathematical expression from the current stress operating region, to V_g corresponding to this region, the V_{gBD} from this region is obtained and then plotted on a log-log scale in Figure 12 for comparison with the linear t_{BD} model from Figure 11 represented by the black dashed line. It is observed that there is a distinct demarcation in the current stress operating region. To the left of this demarcation point, the extracted V_{gBD} is overestimated due to hole trapping, while to the right, the extracted V_{gBD} starts to perfectly match the linear relationship of V_{gBD} measured in DUTs. This strongly validates the feasibility of the mathematical expression for ΔV_{gBD} established for the thermally grown gate oxide in commercial SiC DUTs in previous works. It also confirms that t_{BD} under CCS for DUTs, following a $1/I$ model, is correct and theoretically founded. Therefore, based on Equations (6) and (15), the mathematical model for the Q_{BD} of gate oxide in DUTs under CCS can be established and expressed as follows:

$$Q_{BD}(I) = 0.071 \cdot I^{-0.017} \cdot \mathrm{C} \quad (19)$$

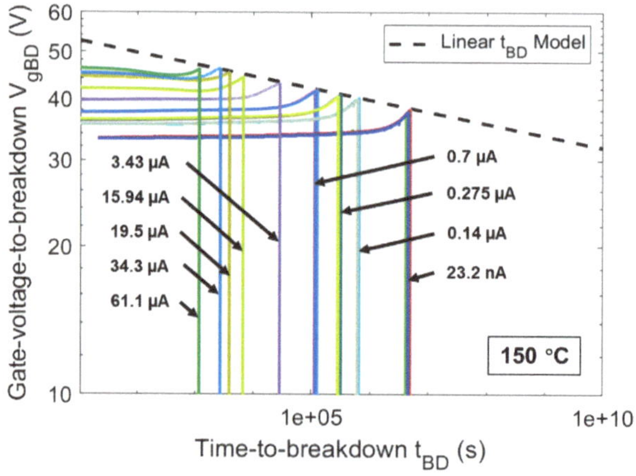

Figure 11. V_{gBD} extracted from V_g curves at 150 °C fitted by a linear t_{BD} model.

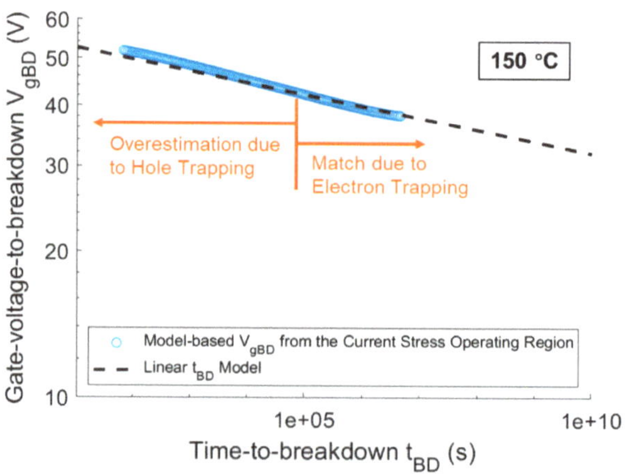

Figure 12. Comparison of model-based V_{gBD} from the current stress operating region at 150 °C with the linear t_{BD} model.

From Equation (19), it can be observed that the power exponent of $1/I$ is 0.017, which approaches zero, causing the power in the expression to be minimally influenced by I and tending toward 1. Consequently, this makes the $Q_{BD}(I)$ for DUTs approach a constant value of 0.071 C, with the influence of I being almost negligible. This is consistent with the failure mechanism of charge-driven breakdown, theoretically supporting the notion that the failure mechanism of thermally grown SiO_2 on SiC under CCS is charge-driven breakdown.

3.3. Extraction of Q_{BD} in Commercial SiC DUTs under CVS and PVS

From Section 3.2, the Q_{BD} model for the gate oxide of DUTs in this work has been established. However, this model has limitations as it is based on the condition of CCS as the stress method for the gate oxide of DUTs. To prove the universality of the model and eliminate the limitations, it is necessary to expand the stress method for the gate oxide of DUTs. CVS, a routine stress method used in the industry for the TDDB test of thermal oxide in commercial SiC power devices, is considered. Additionally, PVS, which more closely replicates the dynamic stress experienced by the thermal oxide in actual operations of commercial SiC power devices, is also taken into account. Figures 13a and 13b respectively show I_{gss} over stress time until the oxide breakdown at 150 °C for DUTs under various

CVS and PVS, with different CVS and PVS scenarios also detailed in the figures. Following Equation (6), Q_{BD} values for DUTs under these two stress methods are extracted and presented in Figure 14. As for the Q_{BD} values for DUTs under CCS, they can be easily extracted through the product of constant I and t_{BD}, depicted in Figure 14 as well. For comparison, the mathematical-model-based Q_{BD} under CCS is also displayed in Figure 14.

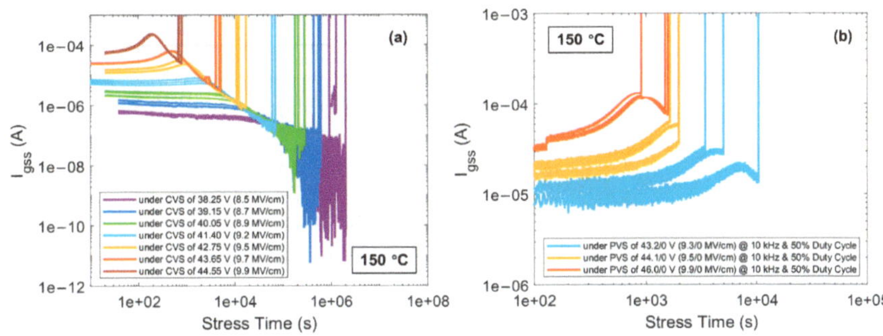

Figure 13. I_{gss} curves as a function of stress time until the oxide breakdown at 150 °C for DUTs under various (**a**) CVS; (**b**) PVS.

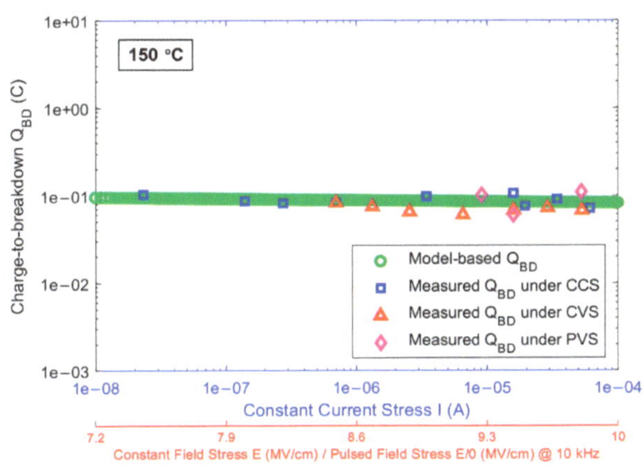

Figure 14. Comparison of model-based Q_{BD} with measured Q_{BD} under CCS, CVS, and PVS at 150 °C.

4. Discussion

From Figure 14, it can be observed that at 150 °C, CVS and PVS correspond to each other through the electric field stress applied on the gate oxide of DUTs. According to the details in Figure 13 for CVS and PVS scenarios, the difference lies in that under CVS, the gate oxide of DUTs is subjected to a continuous electric field stress until the gate oxide breakdown, whereas under PVS, the same electric field stress applied to the gate oxide of DUTs is a pulsed stress with a frequency of 10 kHz and a duty cycle of 50% until the gate oxide breakdown. The electric field stress applied to the gate oxide is roughly estimated by the ratio of the positive voltage applied to the gate and the gate oxide thickness. Under CCS, the gates of DUTs are subjected to a continuous current stress towards the gate oxide until its breakdown, and CCS corresponds to the electric field stress on the gate oxide under V_g associated with the current stress in Figure 1, further corresponding to CVS and PVS. The Q_{BD} values of gate oxide in DUTs extracted under the three different stress methods are distributed in the figure according to the above correspondence and are compared with the model-based Q_{BD} extracted from the Q_{BD} model of thermal gate oxide in DUTs established under CCS. It is significantly observed that the Q_{BD} values of thermal gate oxide in DUTs extracted under the three stress methods conform to the model expectation. The slight differences in the extracted Q_{BD} data fall within the error margin caused by individual differences among the DUTs, which is acceptable and can be almost neglected.

5. Conclusions

In this work, the mathematical model established for describing the electron trapping phenomenon in thermal oxide grown on Si, intended for very thin SiO_2, is considered for transplantation to the gate oxide of commercial SiC power MOSFETs, which is thermally grown on SiC. Given that the mathematical model was initially proven to be applicable only for SiO_2 grown on Si with a thickness not exceeding 10 nm, its applicability to SiO_2 thermally grown on SiC, which is approximately 4–5 times thicker in commercial SiC power MOSFETs, is worth discussing. Based on the CCS-TDDB data of the commercial SiC DUTs featuring approximately 45 nm thick sections of thermal gate oxide, the feasibility of this electron trapping model, under conditions where the oxide charge trapping mechanism is predominantly governed by electron trapping, is confirmed in the commercial SiC cases. Following this model, a Q_{BD} model for the thermal gate oxide of commercial SiC DUTs under CCS is established in this work. Apart from the CCS-TDDB test, the CVS-TDDB and PVS-TDDB tests are also conducted on these DUTs. The Q_{BD} values of thermal gate oxide in DUTs are extracted from the TDDB data under the three different stress methods through the integral of I_{gss} over stress time, and are compared with the established Q_{BD} model. The results demonstrate that the measured Q_{BD} values align with the model expectation, indicating that Q_{BD}, as a characteristic value of the quality of thermal oxide grown on SiC, remains stable and unaffected by the stressors. This is consistent with and confirms the expectation that the failure mechanism of thermal oxide grown on SiC is charge-driven breakdown. This provides a solid theoretical foundation for establishing a new, more accurate lifetime prediction model based on Q_{BD} for commercial SiC power MOSFETs with thermal gate oxide. Additionally, since Q_{BD} is not affected by the stressors and considering the reduced efficiency in extracting Q_{BD} due to the suppression effect of trapped electrons on I_{gss} under CVS, CCS is recommended as a faster and more accurate method for extracting Q_{BD} in the industry, compared with the conventional CVS, for establishing lifetime prediction models based on Q_{BD} for SiC power MOSFETs with thermal gate oxide.

Author Contributions: Conceptualization, J.Q., M.H.W. and A.K.A.; methodology, J.Q. and M.J.; software, J.Q. and L.S.; validation, J.Q., L.S., M.J. and M.B.; formal analysis, J.Q., L.S., M.J., M.B., A.S., H.Y., S.H., M.H.W. and A.K.A.; investigation, J.Q., L.S., M.J., M.B. and H.Y.; resources, J.Q.; data curation, J.Q. and L.S.; writing—original draft preparation, J.Q.; writing—review and editing, J.Q., A.S. and A.K.A.; visualization, J.Q.; supervision, A.K.A.; project administration, A.S.; funding acquisition, A.S. All authors have read and agreed to the published version of the manuscript.

Funding: This research was funded by Ford Auto Co., grant number GR123387.

Institutional Review Board Statement: Not applicable.

Informed Consent Statement: Not applicable.

Data Availability Statement: Data are contained within the article.

Acknowledgments: This research is supported under the Ford-OSU Alliance Project-Phase II to The Ohio State University. The authors would like to thank the team members from Ford Motor Co. under the project for the helpful discussion.

Conflicts of Interest: Author Atsushi Shimbori was employed by the company Ford Motor Co. The remaining authors declare that the research was conducted in the absence of any commercial or financial relationships that could be construed as potential conflicts of interest. The authors declare that this study received funding from Ford Auto Co. The funders had no role in the design of the study; in the collection, analyses, or interpretation of data; in the writing of the manuscript; or in the decision to publish the results.

References

1. Choi, H. *Overview of Silicon Carbide Power Devices*; Fairchild Semiconductor: Sunnyvale, CA, USA, 2016.
2. FastSiC. SiC Is Replacing Si in Electric Vehicles. Available online: https://fastsic.com/2020/06/24/sic-is-replacing-si-in-electric-vehicles/ (accessed on 1 June 2020).
3. Emilio, M. SiC enabling EV applications. Available online: https://www.powerelectronicsnews.com/sic-enabling-ev-applications/ (accessed on 12 April 2019).
4. Russell, S.; Gammon, P. ROHM Gen 4: A Technical Review. Available online: https://www.techinsights.com/blog/rohm-gen-4-technical-review (accessed on 2 August 2022).
5. Telford, M. SiC Making Further Inroads into Silicon for EV Powertrains. *SemiconductorTODAY* **2023**, *18*. Available online: https://www.semiconductor-today.com/news_items/backissues/semiconductor-today-april-2023.pdf (accessed on 17 April 2023).
6. Li, H.F.; Dimitrijev, S.; Harrison, H.B.; Sweatman, D. Interfacial characteristics of N_2O and NO nitrided SiO_2 grown on SiC by rapid thermal processing. *Appl. Phys. Lett.* **1997**, *70*, 2028–2030. [CrossRef]
7. Li, H.; Dimitrijev, S.; Harrison, H.B. Improved reliability of NO-nitrided SiO_2 grown on p-type 4H-SiC. *IEEE Electron Device Lett.* **1998**, *19*, 279–281.
8. Fukuda, K.; Suzuki, S.; Tanaka, T.; Arai, K. Reduction of interface-state density in 4H–SiC n-type metal–oxide–semiconductor structures using high-temperature hydrogen annealing. *Appl. Phys. Lett.* **2000**, *76*, 1585–1587. [CrossRef]
9. Chung, G.Y.; Tin, C.C.; Williams, J.R.; McDonald, K.; Di Ventra, M.; Pantelides, S.T.; Feldman, L.C.; Weller, R.A. Effect of nitric oxide annealing on the interface trap densities near the band edges in the 4H polytype of silicon carbide. *Appl. Phys. Lett.* **2000**, *76*, 1713–1715. [CrossRef]
10. Itoh, H.; Enokizono, T.; Miyase, T.; Hori, T.; Wada, K.; Furumai, M. High-Quality SiC Epitaxial Wafer "EpiEra" Realizing High-Reliability Large-Current Power Devices. *Sumitomo Electr.* **2020**, *91*, 47–50.
11. Zhao, Z.; Li, Y.; Xia, X.; Wang, Y.; Zhou, P.; Li, Z. Growth of high-quality 4H-SiC epitaxial layers on 4° off-axis C-face 4H-SiC substrates. *J. Cryst. Growth* **2020**, *531*, 125355. [CrossRef]
12. Kim, J.; Kim, K. 4H-SiC Double-Trench MOSFET with Side Wall Heterojunction Diode for Enhanced Reverse Recovery Performance. *Energies* **2020**, *13*, 4602. [CrossRef]
13. Chaturvedi, M.; Dimitrijev, S.; Haasmann, D.; Moghadam, H.A.; Pande, P.; Jadli, U. Comparison of Commercial Planar and Trench SiC MOSFETs by Electrical Characterization of Performance-Degrading Near-Interface Traps. *IEEE Trans. Electron Devices* **2022**, *69*, 6225–6230. [CrossRef]
14. Seok, O.; Kang, I.; Moon, J.; Kim, H.; Ha, M.; Bahng, W. Double p-base structure for 1.2-kV SiC trench MOSFETs with the suppression of electric-field crowding at gate oxide. *Microelectron. Eng.* **2020**, *225*, 111280. [CrossRef]
15. Park, Y.; Yoon, H.; Kim, C.; Kim, G.; Kang, G.; Seok, O.; Ha, M. Design and Optimization of 1.2 kV SiC Trench MOSFETs Using a Tilted Ion Implantation Process for High Breakdown Voltage. *Jpn. J. Appl. Phys.* **2023**, *62*, 011001. [CrossRef]
16. Stahlbush, R.; Mahakik, K.; Lelis, A.; Green, R. Effects of Basal Plane Dislocations on SiC Power Device Reliability. In Proceedings of the 2018 IEEE 64th International Electron Devices Meeting (IEDM), San Francisco, CA, USA, 1–5 December 2018.
17. Qian, J.; Shi, L.; Jin, M.; Bhattacharya, M.; Shimbori, A.; Yu, H.; Houshmand, S.; White, M.H.; Agarwal, A.K. An Investigation of Body Diode Reliability in Commercial 1.2 kV SiC Power MOSFETs with Planar and Trench Structures. *Micromachines* **2024**, *15*, 177. [CrossRef] [PubMed]
18. Roccaforte, F.; Greco, G.; Fiorenza, P. Processing Issues in SiC and GaN Power Devices Technology: The Cases of 4H-SiC Planar MOSFET and Recessed Hybrid GaN MISHEMT. In Proceedings of the 2018 International Semiconductor Conference (CAS), Sinaia, Romania, 10–12 October 2018.
19. Yao, J. Working principle and characteristic analysis of SiC MOSFET. *J. Phys. Conf. Ser.* **2023**, *2435*, 012022. [CrossRef]
20. Keukeleire, C. Silicon Carbide (SiC)—From Challenging Material to Robust Reliability. Available online: https://www.onsemi.com/pub/collateral/tnd6396-d.pdf (accessed on 10 December 2022).
21. Ni, Z.; Lyu, X.; Yadav, O.; Singh, B.; Zheng, S.; Cao, D. Overview of Real-Time Lifetime Prediction and Extension for SiC Power Converters. *IEEE Trans. Power Electron.* **2020**, *35*, 7765–7794. [CrossRef]
22. Ikpe, S.; Lauenstein, J.; Carr, G.; Hunter, D.; Ludwig, L.; Wood, W.; Castillo, L.; Fitzpatrick, F.; Chen, Y. Silicon-Carbide Power MOSFET Performance in High Efficiency Boost Power Processing Unit for Extreme Environments. *Addit. Pap. Present.* **2016**, *2016*, 184–189. [CrossRef]
23. Liu, T.; Zhu, S.; White, M.H.; Salemi, A.; Sheridan, D.; Agarwal, A.K. Time-Dependent Dielectric Breakdown of Commercial 1.2 kV 4H-SiC Power MOSFETs. *J. Electron Devices Soc.* **2021**, *9*, 633–639. [CrossRef]

24. McPherson, J.W.; Mogul, H.C. Underlying Physics of the Thermochemical E Model in Describing Low-field Time-dependent Dielectric Breakdown in SiO_2 Thin Films. *J. Appl. Phys.* **1998**, *84*, 1513–1523. [CrossRef]
25. Liu, T. Gate Oxide Reliability of 4H-SiC MOSFETs. Level of Thesis, The Ohio State University, Columbus, OH, USA, 11 April 2022.
26. Ghetti, A. Gate Oxide Reliability: Physical and Computational Models. Available online: https://link.springer.com/chapter/10.1007/978-3-662-09432-7_6 (accessed on 8 May 2004).
27. Moens, P.; Franchi, J.; Lettens, J.; Schepper, L.D.; Domeij, M.; Allerstam, F. A Charge-to-Breakdown (QBD) Approach to SiC Gate Oxide Lifetime Extraction and Modeling. In Proceedings of the 2020 32nd International Symposium on Power Semiconductor Devices and ICs (ISPSD), Vienna, Austria, 13–18 September 2020.
28. Tan, W.; Zhao, L.; Lu, C.; Nie, W.; Gu, X. An In-depth Investigation of Gate Leakage Current Degradation Mechanisms in 1.2kV 4H-SiC Power MOSFETs. *Microelectron. Reliab.* **2023**, *142*, 114907. [CrossRef]
29. Liang, M.; Hu, C. Electron trapping in very thin thermal silicon dioxides. In Proceedings of the 1981 International Electron Devices Meeting (IEDM), Washington, DC, USA, 7–9 December 1981.
30. Wang, H.; Jiang, D. Design of High Temperature Gate Driver for SiC MOSFET for EV Motor Drives. In Proceedings of the 2017 IEEE Transportation Electrification Conference and Expo, Asia-Pacific (ITEC Asia-Pacific), Harbin, China, 7–10 August 2017.
31. Zhu, S.; Liu, T.; Shi, L.; Jin, M.; Maddi, H.; White, M.H.; Agarwal, A.K. Comparison of Gate Oxide Lifetime Predictions with Charge-to-Breakdown Approach and Constant-Voltage TDDB on SiC Power MOSFET. In Proceedings of the 2021 IEEE 8th Workshop on Wide Bandgap Power Devices and Applications (WiPDA), Redondo Beach, CA, USA, 7–11 November 2021.
32. Shi, L.; Zhu, S.; Qian, J.; Jin, M.; Bhattacharya, M.; Shimbori, A.; Liu, T.; White, M.H.; Agarwal, A.K. Investigation of different screening methods on threshold voltage and gate oxide lifetime of SiC Power MOSFETs. In Proceedings of the 2023 IEEE International Reliability Physics Symposium (IRPS), Monterey, CA, USA, 26–30 March 2023.
33. Sarnago, H.; Lucía, Ó.; Jiménez, R.; Gaona, P. Differential-Power-Processing On-Board-Charger for 400/800-V Battery Architectures using 650-V Super Junction MOSFETs. In Proceedings of the 2021 IEEE Applied Power Electronics Conference and Exposition (APEC), Phoenix, AZ, USA, 14–17 June 2021.
34. Kegley, L. Exploring On-Board EV Systems from 400 V to 800 V. Available online: https://www.powerelectronicsnews.com/exploring-on-board-ev-systems-from-400-v-to-800-v/ (accessed on 17 April 2023).
35. Jung, C. Power Up with 800-V Systems: The benefits of upgrading voltage power for battery-electric passenger vehicles. *IEEE Electrif. Mag.* **2017**, *5*, 53–58. [CrossRef]
36. Hussey, A. Lucid Air to be the Fastest Charging EV, Featuring a 900V+ Architecture Delivering a Charging Rate of Up to 20 Miles Per Minute. Available online: https://lucidmotors.com/media-room/lucid-air-fastest-charging-ev (accessed on 19 August 2020).
37. Goldberg, L. Exploiting SiC MOSFETs to Power EV Innovation. Available online: https://www.electronicdesign.com/markets/automotive/article/21262547/electronic-design-exploiting-sic-mosfets-to-power-ev-innovation (accessed on 23 March 2023).
38. Shi, L.; Qian, J.; Jin, M.; Bhattacharya, M.; Yu, H.; White, M.H.; Agarwal, A.K.; Shimbori, A.; Xu, Z. An Effective Screening Technique for Early Oxide Failure in SiC Power MOSFETs. In Proceedings of the 2023 IEEE 10th Workshop on Wide Bandgap Power Devices & Applications (WiPDA), Charlotte, NC, USA, 4–6 December 2023.
39. Koga, Y.; Kurita, K. Fabrication of Silicon on Insulator Wafer with Silicon Carbide Insulator Layer by Surface-activated Bonding at Room Temperature. *Jpn. J. Appl. Phys.* **2020**, *59*, 051002. [CrossRef]
40. Pu, S. Reliability Assessment, Condition Monitoring and Lifetime Estimation of Silicon Carbide MOSFETs. Ph.D. Thesis, The University of Texas at Dallas, Dallas, TX, USA, 20 July 2021.

materials

MDPI

Article

Study of Carrier Mobilities in 4H-SiC MOSFETS Using Hall Analysis

Suman Das [1,*], Yongju Zheng [1,2], Ayayi Ahyi [1], Marcelo A. Kuroda [1] and Sarit Dhar [1,*]

1 Department of Physics, Auburn University, Auburn, AL 36849, USA
2 SemiQ Inc., Lake Forest, CA 92630, USA
* Correspondence: szd0073@auburn.edu (S.D.); szd0024@auburn.edu (S.D.)

Abstract: The channel conduction in 4H-SiC metal–oxide–semiconductor field effect transistors (MOSFETs) are highly impacted by charge trapping and scattering at the interface. Even though nitridation reduces the interface trap density, scattering still plays a crucial role in increasing the channel resistance in these transistors. In this work, the dominant scattering mechanisms are distinguished for inversion layer electrons and holes using temperature and body-bias-dependent Hall measurements on nitrided lateral 4H-SiC MOSFETs. The effect of the transverse electric field (E_{eff}) on carrier mobility is analyzed under strong inversion condition where surface roughness scattering becomes prevalent. Power law dependencies of the electron and hole Hall mobility for surface roughness scattering are determined to be $E_{eff}^{-1.8}$ and $E_{eff}^{-2.4}$, respectively, analogous to those of silicon MOSFETs. Moreover, for n-channel MOSFETs, the effect of phonon scattering is observed at zero body bias, whereas in p-channel MOSFETs, it is observed only under negative body biases. Along with the identification of regimes governed by different scattering mechanisms, these results highlight the importance of the selection of substrate doping and of E_{eff} in controlling the value of channel mobility in 4H-SiC MOSFETs.

Keywords: 4H-SiC MOSFET; nitridation; scattering; Hall measurements; body bias; transverse electric field

Citation: Das, S.; Zheng, Y.; Ahyi, A.; Kuroda, M.A.; Dhar, S. Study of Carrier Mobilities in 4H-SiC MOSFETS Using Hall Analysis. *Materials* **2022**, *15*, 6736. https://doi.org/10.3390/ma15196736

Academic Editors: Marilena Vivona and Mike Jennings

Received: 2 September 2022
Accepted: 25 September 2022
Published: 28 September 2022

1. Introduction

Silicon carbide (4H-SiC) is one of the primary wide-band-gap semiconductors for high power and harsh environment applications because of its physical properties, such as a high critical electric field and high thermal conductivity [1]. Discrete 4H-SiC diodes and metal–oxide–semiconductor field effect transistors (MOSFETs) are being widely adopted for high voltage power conversion in hybrid/electric vehicles [2], solar and wind energy generation [3], and various high temperature applications [4], enabling significant advances for next-generation energy efficient power systems. Moreover, 4H-SiC is a unique candidate for the development of an integrated circuit (IC) technology that is still in its infancy. 4H-SiC IC technology operating at very high temperatures >300 °C [5,6] is very attractive, as it enables operation in environments and ambientes that are not accessible to conventional silicon or silicon on insulator (SOI) platforms [7,8]. Lateral complementary metal–oxide–semiconductor (CMOS) IC technology in 4H-SiC is desirable for the fabrication of large-scale integration devices due to its high noise immunity and low static power consumption [9]. To be materialized, such technology demands both n- and p-channel MOSFETs capable of operating at high-temperatures. While n-channel MOSFETs have reached a suitable degree of maturity through the nitridation of the SiO_2/4H-SiC interface [10–12], mechanisms governing electrical transport in their p-channel counterparts must be investigated to reach comparable levels of development in terms of channel conductivity and device stability.

Recent studies [10,13] on 4H-SiC MOSFETs use field-effect mobility models to analyze channel transport. Additionally, research by Mikami et al. [13] explored the importance of body bias and explained how a body bias experiment can provide predictions of transistor characteristics for MOSFETs made with higher or lower substrate doping using field effect mobility on the weak inversion region. However, the channel scattering mechanisms for p-channel 4H-SiC MOSFET remain unexplored using Hall analysis. Hall mobility is more accurate than field effect mobility, as the carrier concentration is measured independently and does not get affected by trapped charges. A comprehensive study for both kinds of carriers will enable control of the operation region and the preferable substrate doping concentrations for high-temperature IC design. To this end, the channel mobility of nitric oxide (NO)-annealed n- and p-channel 4H-SiC MOSFETs in the strong inversion regime was characterized using Hall measurements. Measurements were performed using the body-bias technique, which allows variation of the effective electric field (E_{eff}) in the channel [13,14]. This approach allows one to differentiate and establish the dominant scattering mechanisms limiting carrier transport, in different carrier density and substrate doping regimes. The power law that governs the relation between μ_{Hall} and E_{eff} was determined for both electrons and holes in the surface-roughness-scattering-dominant regime. In addition, it is demonstrated that a combination of temperature and body bias enables observation of the dominant mechanisms transitioning from surface roughness to phonon scattering.

2. Materials and Methods

2.1. Fundamentals on Channel-Transverse Electric Fields and Mobility

The channel's resistance depends on the channel's inversion-carrier concentration and its mobility (μ_{Hall}), which are related to the atomic-scale composition, structure, and defects at the SiO_2/4H-SiC interface. The mobile carrier's concentration is reduced by traps at electrically-active defect sites, and the channel's mobility is degraded by the interfacial scattering processes. When the MOSFFET is on, the dominant scattering mechanisms at a SiO_2/4H-SiC interface that limits μ_{Hall} are: (i) Coulomb scattering occurring due to trapped charges present at or near the interface oxide and from ionized impurities in the depletion region close to the channel; (ii) phonon scattering occurring due to the interaction of the carriers with lattice vibration in the channel and the surface; and (iii) surface roughness due to imperfections of the 4H-SiC interface [15–17]. The resulting carrier mobility is given by Matthiesen's rule [18]:

$$\frac{1}{\mu_{Hall}} = \frac{1}{\mu_C} + \frac{1}{\mu_{ph}} + \frac{1}{\mu_{SR}} \quad (1)$$

Here, μ_C, μ_{ph}, and μ_{SR} are the Coulomb, phonon, and surface roughness scattering limited mobilities, respectively.

On Si MOSFETs [19,20], earlier studies suggest that the transverse electric field E_{eff} is an important parameter for influencing scattering in the channel. The magnitude of E_{eff} can be obtained as [19],

$$E_{eff} = \frac{1}{\epsilon}\left(\sqrt{2N_A q\epsilon.(2\phi_B \pm V_{BS})} + \eta q n_s\right) \quad (2)$$

The first term in Equation (2) represents the amount of depletion charge, and the second term stands for the induced inversion layer charge excluding trapped charges. Here the "+" sign is applicable for n-channel MOSFETs and the "−" sign is for p-channel MOSFETs. A schematic diagram of these two charge distributions is shown in the left panel of Figure 1. In Equation (2), η is a constant with a value of 0.5 for n-channel and 0.33 for p-channel Si MOSFETs at 27 °C [19–22], which are considered the same for 4H-SiC [13,14,23,24], but this choice does not affect the results and may need further studies in a wider effective field range. ϵ is the dielectric constant, q is the electron charge, n_s is the free carrier concentration, and N_A is the p-well substrate doping concentration for n-channel MOSFETs. V_{BS} is the substrate or body to source bias. $\phi_B = \frac{KT}{q} \ln \frac{N_A}{n_i}$ is the bulk potential. K is the Boltzmann constant and n_i denotes the temperature-dependent intrinsic carrier concentration. Note, for p-channel MOSFET, the n-well substrate doping concentration is denoted by N_D. Application of body/substrate bias acts as a separate knob to control the depletion width without affecting the channel carrier concentration. A variation of channel electric field to study its effect on the conduction and the effects of substrate doping concentration is possible through V_{BS}.

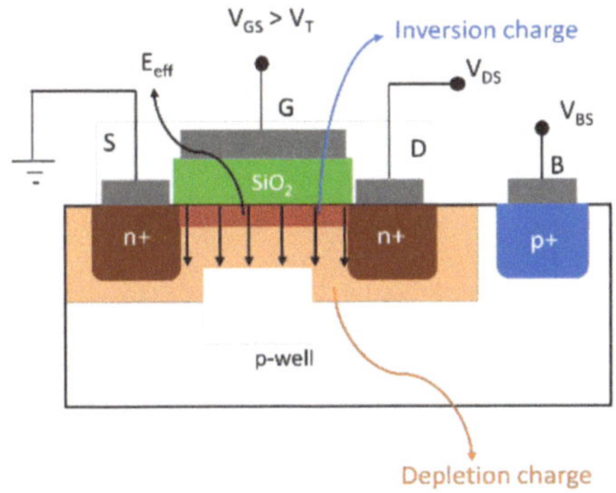

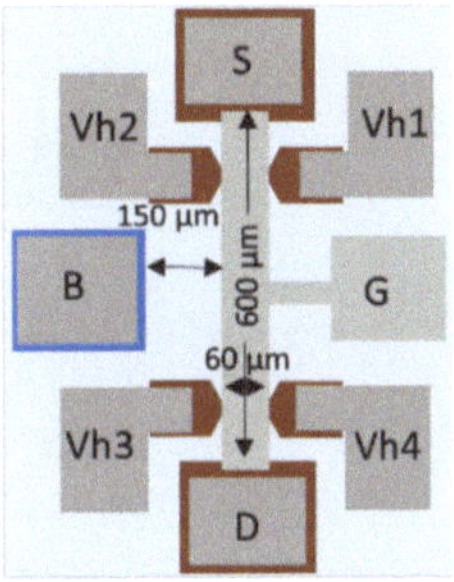

Figure 1. (**left**) Schematic diagram of lateral n-channel Hall MOSFET with the transverse electric field E_{eff} shown, which depends on space charge and inversion charge in the channel at a gate voltage higher than the threshold voltage. (**right**) Top view of a schematic of the fabricated Hall bar MOSFET. Vh1, Vh2, Vh3, and Vh4 represent the Hall voltage terminals.

2.2. Experimental Methods

Hall bar MOSFETs with a channel length of 600 µm and a channel width of 60 µm were fabricated on a 4° off-axis (0001), Si-face oriented 4H-SiC substrate with a p-type epilayer ($N_A - N_D = 6 \times 10^{15}$ cm^{-3}) for n-channel MOSFETs and an n-type epilayer ($N_D - N_A = 6.2 \times 10^{15}$ cm^{-3}) for p-channel MOSFETs. Additional to the Hall bar MOSFETs, T-gated MOSFETs were also present in the chip with a 200 µm × 200 µm channel (width, length). The cross-sectional schematic of the Hall bar device is shown in Figure 1 (left). A nominally uniform doping profile was created using nitrogen and aluminum ion implantations at 700 °C to form n+ and p+ layers on the MOSFETs, respectively, to form source, drain and body contacts. The implanted layers were then activated by annealing at 1650 °C for 30 min in flowing Ar with a graphitic cap to protect the surface. After removal of the cap, gate oxidation was carried out at 1150 °C for 10 h followed by post oxidation annealing at 1175 °C for 2 h in NO. The thickness of the gate oxide was measured to be about ~60 nm and 55 nm by the capacitance voltage method (CV) for n- and p-channel MOSFETs, respectively. Contact photolithography and reactive ion etching were carried out to remove the oxide from the source/drain/body (S/D/B) regions and Hall voltage (V_h) terminals. For n-channel MOSFET, Al was evaporated for all the contacts and annealed at 800 °C for 30 min in flowing Ar. On the other hand, for p-channel MOSFET, Ti was sputtered and lifted off to define the S/D/B/V_h contacts, as shown in Figure 1 (right), following which annealing at 1000 °C for 2 min was performed in flowing Ar to form ohmic contacts. Subsequently, aluminum (55 nm) was thermally evaporated as the gate metal and patterned using lift-off. An overlayer of Au/Cr was sputtered on all the contacts of both kinds of MOSFETs. Finally, the Hall bar MOSFET was mounted on a ceramic chip with epoxy and wire bonded to gold pads on the chip.

3. Results and Discussions

3.1. Classification of the Dominant Scattering Mechanisms

After fabrication of the 4H-SiC MOSFETs, transfer characteristics were obtained at various temperatures. The temperature for the n-channel MOSFETs was varied from 77 to 373 K. For the p-channel MOSFETs, on the other hand, the temperature was raised above room temperature (300 K) until 573 K. The absolute value of threshold voltage became smaller at higher temperatures, as shown in Figure 2. This is associated with the reduction of occupied interface trap densities (D_{it}) at higher temperatures [25]. In addition, the field effect mobility for electrons ($\mu_{fe,\ n}$) was observed to increase from 77 to 296 K. For the p-channel MOSFET, the field effect mobility of holes was observed to be weakly dependent on temperature under strong inversion.

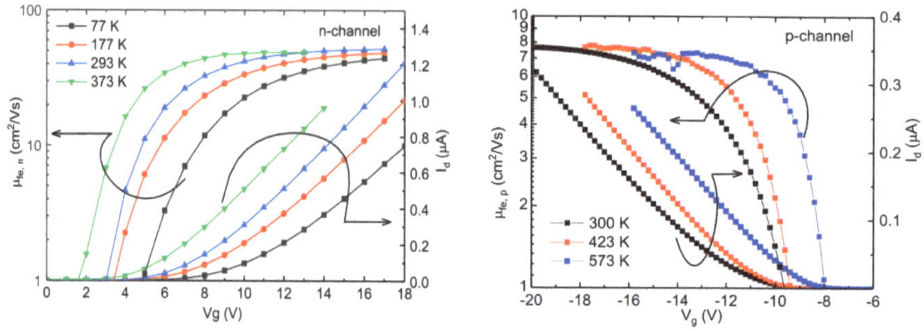

Figure 2. Field effect mobility of electrons and holes versus gate voltage at different temperatures for n- and p-channel 4H-SiC MOSFETs. For n-channel MOSFETs, channel width/length was 60 µm/600 µm, oxide thickness was 60 nm, V_{ds} was set to 0.4 V, and temperature was varied from 77 to 373 K. For p-channel MOSFETs, channel width/length was 200 µm/200 µm, oxide thickness was 55 nm, V_{ds} was set to 0.1 V, and temperature was varied from 300 to 573 K.

To confirm the dominant scattering mechanisms in these devices, Hall measurements were carried out under a perpendicular magnetic field of 0.6 T. At a high gate voltage overdrive, the Hall and field effect mobilities merge, and at low and intermediate gate voltages, field effect mobilities are lower than Hall mobilities due to the presence of trapped charges (see supplementary material). Figure 3 shows μ_{Hall} versus carrier concentration curves at different temperatures and at zero body bias for electrons and holes in n- and p-channel 4H-SiC MOSFETs. To obtain μ_{Hall} versus carrier concentration plot, the gate to source voltage V_g was varied from +2 to +10 V for n-channel and −6 to −20 V for p-channel MOSFETs at constant drain to source voltage $V_{DS} = 0.75$ V. The threshold voltage was found to be +5 V for electrons (at 293 K) and −10 V for holes (300 K), based on linear extrapolation of the I_D–V_G characteristics (Figure 2). For n-channel MOSFETs, Hall mobility above room temperature drops. An increase in mobility from 77 to 293 K is a signature of prevalent Coulomb scattering, whereas above 293 K, mobility decreases with increasing temperature when phonon scattering is dominant. Therefore, for electrons, transport is limited by Coulomb and phonon scattering. From the trend of the experimental results in figure in Section 3.3 and other published results [23,26], electron mobility will reduce below 77 K due to increased Coulomb scattering. Above 373 K, mobility would also decrease, but due to increased phonon scattering. The presence of surface roughness scattering at higher concentration can also be observed in earlier reports [14]. Conversely, for holes in p-channel MOSFET, Hall mobility is seen to increase at lower carrier concentration (for approximately $p_s < 6 \times 10^{11}$ cm^{-2} at 300 K), signifying dominant Coulomb scattering, and becomes independent of temperature at approximately $p_s = 10^{12}$ cm^{-2} (see supplementary materials). Temperature-independent mobility is a sign of surface roughness scattering. Therefore, hole mobility in 4H-SiC p-channel MOSFET is limited by Coulomb scattering at weak inversion and dominant surface roughness scattering at strong inversion.

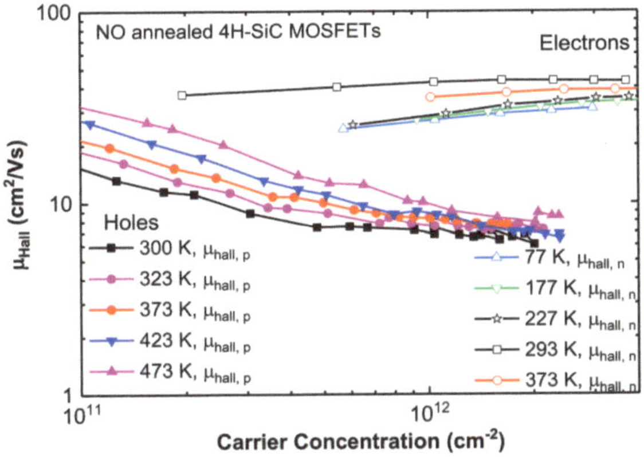

Figure 3. Typical Hall mobility as a function of carrier concentration at different temperatures for n- and p-channel MOSFETs with channel width 60 μm and channel length 600 μm for both kinds of Hall bar MOSFET, and V_{ds} was set to 0.75 V.

After distinguishing the dominant scattering mechanisms in these devices, Hall measurements were carried out under the effect of body bias (V_{BS}) so that the Hall mobility (μ_{Hall}) can be studied as a function of transverse electric field $\left(E_{eff}\right)$ in the strong inversion regime. Recent studies [13,14] observed the effect of transverse electric field on mobility in the strong inversion regime maintaining a fixed carrier concentration. A constant carrier concentration realizes fixed screening from scattering centers and enables the assessment of the sole effect of surface roughness scattering on mobility.

3.2. Analysis of Surface Roughness Scattering Using Body Bias Measurements

First, to confirm the body bias effect on each transistor, a transfer characteristic at different body bias levels was measured, as shown in Figure 4. The threshold voltages (V_{th}) at 300 K and V_{BS} = 0 were approximately +4 V and −10 V for n- and p-channel MOSFETs respectively, based on linear extrapolation of the I_D–V_G characteristics. It can be observed that the threshold voltage increases as the V_{BS} goes from forward to reverse bias for both types of MOSFETs.

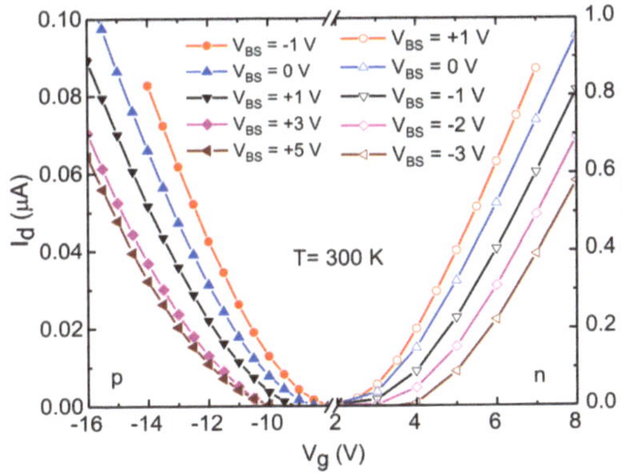

Figure 4. Transfer characteristics at room temperature (300 K) for different body biases (V_{BS}) for n- and p-channel MOSFETs. I_d–V_g sweeps were taken at a constant drain-to-source voltage (0.75 V). A change in V_{th} can be observed due to the modulation of the depletion layer width at the body to source junction with the application of body bias. Note that the y-axis scales are different for the two kinds of MOSFETs.

A positive increment in the V_{BS} widens the depletion layer between the source and the body junction, thereby requiring a higher gate voltage to achieve same level of drain current (I_d) as in lower V_{BS}. The plot of μ_{Hall} as a function of carrier concentration in the strong inversion region for different body biases is shown in Figure 5. An increase in reverse body bias increases the magnitude of the transverse electric field (Equation (2)) and pushes the carriers closer to the surface, making the channel width thinner (~1 nm) at strong inversion.

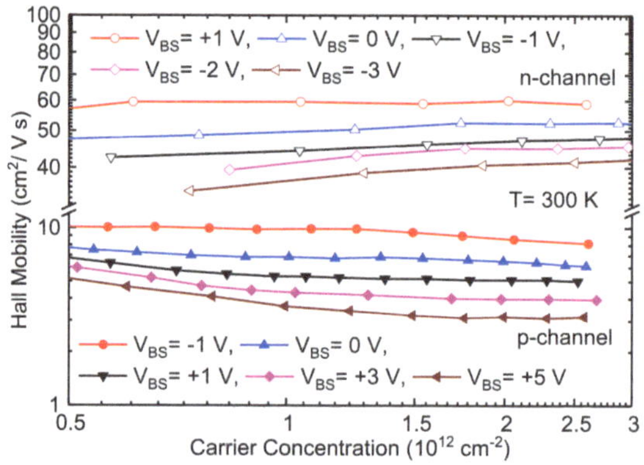

Figure 5. Hall mobility as a function of channel carrier concentration at different body biases at 27 °C. The body bias was varied from +1 to −3 V for n-channel MOSFETs and −1 to +5 V for p-channel MOSFETs; the gate bias V_{GS} was swept to modulate the carrier concentrations under a fixed body bias. A forward body bias value decreases the depletion layer width between source and body, increasing mobility, and at reverse body bias, E_{eff} increases due to expansion of depletion layer, which in turn decreases the channel mobility.

Hence, at a high gate voltage, when mobile holes are closer to the surface, surface roughness scattering becomes even more prevalent through the influence of the varying perturbed potential energy [15]. This is attributed to charge carrier wave functions becoming more susceptible to the fluctuating perturbation near the interface stemming from the crystal miscut (4° off axis substrate) and nano-steps/roughness. As a result, an increased E_{eff} lowers the value of μ_{Hall}. Conversely, a negative body bias decreases the value of E_{eff}, which widens the channel region and lessens the impact of the surface roughness scattering.

These results were used to extract the functional dependence of μ_{Hall} at a fixed carrier concentration to analyze surface roughness scattering under constant screening. For this purpose, using Equation (2), n_s and p_s were converted to E_{eff} for each V_{BS} above 0 V in the strong inversion regime. Next, μ_{Hall} was extracted at fixed carrier concentrations ($1.0 \times 10^{12}, 1.5 \times 10^{12}$, 2.0×10^{12} and 2.5×10^{12} cm^{-2}) and plotted against E_{eff}, as shown in Figure 6. The values of V_{BS} were chosen above zero volts, where the surface roughness becomes the dominant mechanism and there is a minimal contribution from phonon scattering. The curves have been fit using a power law function of E_{eff}, yielding exponents of (-1.8 ± 0.2) and (-2.4 ± 0.3) for channel electrons and holes in 4H-SiC n- and p-channel MOSFETs, respectively. These values are close to those found earlier for surface roughness scattering in Si MOSFETs [20], suggesting that at high E_{eff}, surface imperfections have a similar effect on channel carriers in 4H-SiC as in Si MOSFETs. However, differently from Si, where μ_{Hall} versus E_{eff} curves merge with each other when surface roughness scattering becomes the dominant regime, this universal behavior is not visible for 4H-SiC in Figure 6. This is consistent with earlier studies on 4H-SiC n-channel MOSFETs [14,23,24,27,28]. Perhaps for a single 4H-SiC MOSFET, the measurable range of E_{eff} is not large enough before the breakdown of the gate dielectric to observe the merging of μ_{Hall} at higher

E_{eff}, and a larger range of measurements using devices with different substrate doping is necessary, as reported in [24].

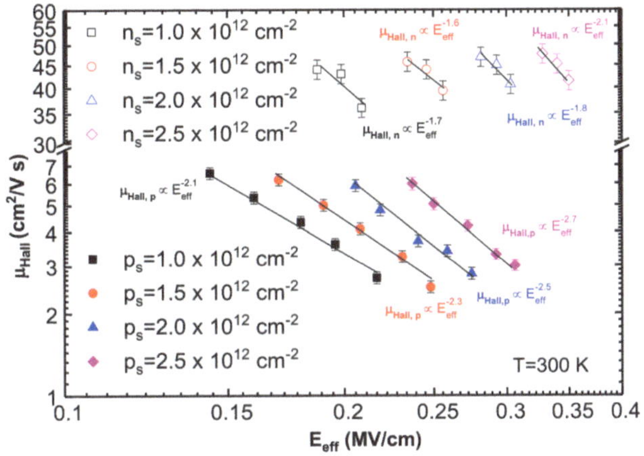

Figure 6. Transverse electric field (E_{eff}) dependence of Hall mobility (μ_{Hall}) at different body biases (−1, −2, and −3 V for n-channel and 0, +1, +3, +5, and +7 V for p-channel) at room temperature (27 °C) for different carrier concentrations (1.0×10^{12}, 1.5×10^{12}, 2.0×10^{12}, and 2.5×10^{12} cm^{-2}), in log–log scale. The different E_{eff} points for a particular curve were obtained from different body biases at a fixed carrier concentration from Figure 5. The average power law that μ_{Hall} follows on E_{eff} is (-1.8 ± 0.2) for n-channel MOSFETs and (-2.4 ± 0.3) for p-channel MOSFETs, providing the power of E_{eff} for surface roughness scattering for electrons and holes in 4H-SiC MOSFETs. Error bars signify ~5% error estimated based on variation in at least 3 measurements.

3.3. *Phonon-Scattering-Limited Mobility*

For n-channel MOSFETs, dropping mobility was observed above room temperature at zero body bias, which signifies the dominance of phonon-scattering-limited mobility, as seen in Figure 7. This fact is consistent with recent work [23] which reported that phonon-scattering-limited mobility can be observed for n-channel MOSFETs fabricated on lightly doped ($\lesssim 5 \times 10^{15}$ cm^{-3}) p-type epitaxial layers. However, for p-channel MOSFETs, phonon-scattering-limited mobility was absent at zero body bias in the measured temperature range. In this case, phonon scattering was visible only when the source to body junction was kept at a forward bias ($V_{BS} < 0$). Then, the power law-dependence of the μ_{Hall} on E_{eff} was observed to give an exponent of −0.32 (see supplementary materials). A power law of $E_{eff}^{-1/3}$ is an indication of phonon scattering [20,23]. In our study, a negative body bias was used to replicate a lightly doped n-type substrate through the "effective doping concentration" [14] of the n-well $N_{D,eff}$, given by:

$$N_{D,eff} = N_D\left(1 + \frac{V_{BS}}{2\phi_B}\right). \tag{3}$$

At a body bias $V_{BS} = -1.3$ V and a fixed $p_s = 10^{12}$ cm^{-2}, a low value of $N_{D,eff}$ (approximately 3×10^{15} cm^{-3}) can be maintained in the 300 to 498 K (27 °C to 225 °C) temperature range. Figure 7 shows that μ_{Hall} increases from room temperature to 398 K due to Coulomb scattering, after which it decreases owing to the dominance of phonon scattering following a power law of $T^{-0.9}$.

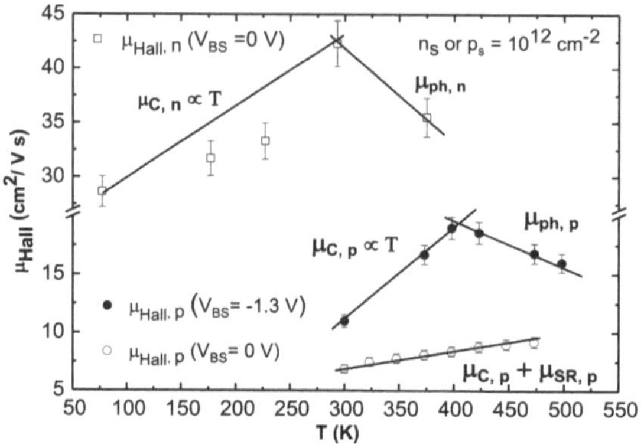

Figure 7. Hall mobility (μ_{Hall}) versus temperature T for different body biases is plotted for a fixed n_s /or p_s = 10^{12} cm^{-2}. For n-channel MOSFET, Coulomb scattering is dominant below RT; above RT phonon scattering can be observed at zero body bias. However, for p-channel MOSFETs, μ_{Hall} is limited by a combined effect of Coulomb and surface roughness scattering, V_{BS} = 0 V. For V_{BS} = −1.3 V, μ_{Hall} increases linearly until 398 K (125 °C) due to Coulomb scattering. At higher temperatures, phonon scattering dominates and μ_{Hall} decreases with temperature as $T^{-0.9}$. Error bars signify variation in in at least 3 measurements.

The power of T found here is near to the theoretical value of T^{-1} [29]. Under negative V_{BS}, the channel is thick enough to interact with the n-well lattice vibrations at $T > 398$ K, and the hole mobility is limited by phonon scattering. This is consistent with n-channel 4H-SiC MOSFETs fabricated on lightly doped substrates, where phonon-scattering-limited mobility at low E_{eff} or $N_{D,eff}$ can be observed [23]. Conversely, for $V_{BS} = 0$ V, channel thickness is comparatively thinner than at $V_{BS} = -1.3$ V. In this case, no photon-limited scattering is observable, but μ_{Hall} is limited by the cumulative effect of Coulomb and surface roughness scattering. As a result, μ_{Hall} increases slightly at higher temperatures.

4. Conclusions

In conclusion, the dominant scattering mechanisms in n- and p- 4H-SiC MOSFET channels were distinguished using Hall measurements. Electron mobility in n-channel MOSFET is limited by Coulomb scattering in weak inversion and a combination of phonon and surface roughness scattering in the strong inversion region. On the other hand, hole mobility in p-channel MOSFETs is limited primarily by Coulomb and surface roughness scattering. Body-bias- and temperature-dependent Hall measurements were performed and analyzed under strong inversion to study mobility in terms of effective transvers electric field (E_{eff}). As body bias is increased, a higher value of E_{eff} confines the mobile carriers near the 4H-SiC/SiO_2 interface, resulting in surface-roughness-scattering-limited mobility. A changing body bias was applied to determine the power-law dependence of channel electron and hole mobility on the E_{eff} for surface roughness scattering in 4H-SiC as $E_{eff}^{-1.8}$ and $E_{eff}^{-2.4}$, respectively. Furthermore, with the application of negative V_{BS}, the depletion layer between the source and body contact decreases, and the resulting low E_{eff} leads to higher mobility at a given gate bias. For p-channel MOSFET, at a forward V_{BS} (−1.3 V at 398 K), the channel is thick enough to interact with the n-well lattice vibration, giving rise to a phonon-scattering-limited hole mobility. Therefore, at low negative V_{BS}, the channel conduction behaves like a MOSFET with a lightly doped substrate. These new findings emphasize the significance of substrate doping selection for 4H-SiC MOSFETs. Additionally, for the development of highly efficient 4H-SiC CMOS devices for high temperature operation, the three types of scattering processes identified here must be considered to depend on the value of the transverse electric field in the channel and the operating temperature.

Supplementary Materials: The following supporting information can be downloaded at https://www.mdpi.com/article/10.3390/ma15196736/s1. Figure S1: Field effect versus Hall mobility. Figure S2: Surface roughness scattering at a high hole concentration. Figure S3: Hole mobility versus transverse electric field for different body biases at room temperature.

Author Contributions: The research was conducted under the supervision of S.D. (Sarit Dhar). The devices were fabricated and characterized by S.D. (Suman Das) and Y.Z. The data were analyzed, and the manuscript was prepared by S.D. (Suman Das). A.A. helped in fabrication and measurements. M.A.K. and S.D. (Sarit Dhar) reviewed and improved the manuscript. All authors have read and agreed to the published version of the manuscript.

Funding: This research was partially funded by the US Department of Energy (subcontracted to National Renewable Energy Laboratory, grant NREL-AHL-9-92632-01).

Institutional Review Board Statement: Not applicable.

Informed Consent Statement: Not applicable.

Data Availability Statement: The data that support the findings of this study are available from the corresponding authors upon reasonable request.

Acknowledgments: The authors acknowledge the support of Tamara Isaacs-Smith of Auburn University for help with device fabrication.

Conflicts of Interest: The authors do not have any potential conflict of interest to disclose.

References

1. Baliga, B.J. *Fundamentals of Power Semiconductor Devices*; Springer Science & Business Media: Berlin, Germany, 2010; ISBN 978-0-387-47314-7.
2. Agarwal, A.K. An Overview of SiC Power Devices. In Proceedings of the 2010 International Conference on Power, Control and Embedded Systems, Allahabad, India, 29 November–1 December 2010; pp. 1–4.
3. He, J.; Zhao, T.; Jing, X.; Demerdash, N.A.O. Application of Wide Bandgap Devices in Renewable Energy Systems—Benefits and Challenges. In Proceedings of the 2014 International Conference on Renewable Energy Research and Application (ICRERA), Milwaukee, WI, USA, 19–22 October 2014; pp. 749–754.
4. Hunter, G.; Neudeck, P.; Xu, J.C.; Lukco, D.; Trunek, A.J.; Artale, M.; Lampard, P.; Androjna, D.; Makel, D.; Ward, B.; et al. Development of SiC-Based Gas Sensors for Aerospace Applications. *MRS Online Proc. Libr.* **2004**, *815*. [CrossRef]
5. Rahman, A.; Roy, S.; Murphree, R.; Kotecha, R.; Addington, K.; Abbasi, A.; Mantooth, H.A.; Francis, A.M.; Holmes, J.; Di, J. High-Temperature SiC CMOS Comparator and Op-Amp for Protection Circuits in Voltage Regulators and Switch-Mode Converters. *IEEE J. Emerg. Sel. Top. Power Electron.* **2016**, *4*, 935–945. [CrossRef]
6. Holmes, J.; Francis, A.M.; Getreu, I.; Barlow, M.; Abbasi, A.; Mantooth, H.A. Extended High-Temperature Operation of Silicon Carbide CMOS Circuits for Venus Surface Application. *J. Microelectron. Electron. Packag.* **2016**, *13*, 143–154. [CrossRef]
7. Kappert, H.; Kordas, N.; Dreiner, S.; Paschen, U.; Kokozinski, R. High Temperature SOI CMOS Technology and Circuit Realization for Applications up to 300 °C. In Proceedings of the 2015 IEEE International Symposium on Circuits and Systems (ISCAS), Lisbon, Portugal, 24–27 May 2015. [CrossRef]
8. Homberger, J.M.; Cilio, E.; Schupbach, R.M.; Lostetter, A.B.; Mantooth, H.A. A High-Temperature Multichip Power Module (MCPM) Inverter Utilizing Silicon Carbide (SiC) and Silicon on Insulator (SOI) Electronics. In Proceedings of the 2006 37th IEEE Power Electronics Specialists Conference, Jeju, Korea, 18–22 June 2006; pp. 1–7.
9. Liu, D.; Svensson, C. Power Consumption Estimation in CMOS VLSI Chips. *IEEE J. Solid-State Circuits* **1994**, *29*, 663–670. [CrossRef]
10. Tachiki, K.; Kimoto, T. Improvement of Both N- and p-Channel Mobilities in 4H-SiC MOSFETs by High-Temperature N_2 Annealing. *IEEE Trans. Electron Devices* **2021**, *68*, 638–644. [CrossRef]
11. Tachiki, K.; Kaneko, M.; Kimoto, T. Mobility Improvement of 4H-SiC (0001) MOSFETs by a Three-Step Process of H_2 Etching, SiO_2 Deposition, and Interface Nitridation. *Appl. Phys. Express* **2021**, *14*, 031001. [CrossRef]
12. Tachiki, K.; Mikami, K.; Ito, K.; Kaneko, M.; Kimoto, T. Mobility Enhancement in Heavily Doped 4H-SiC (0001), $(11\bar{2}0)$, and $(1\bar{1}00)$ MOSFETs via an Oxidation-Minimizing Process. *Appl. Phys. Express* **2022**, *15*, 071001. [CrossRef]
13. Mikami, K.; Tachiki, K.; Ito, K.; Kimoto, T. Body Doping Dependence of Field-Effect Mobility in Both n- and p-Channel 4H-SiC Metal-Oxide-Semiconductor Field-Effect Transistors with Nitrided Gate Oxides. *Appl. Phys. Express* **2022**, *15*, 036503. [CrossRef]

14. Hirai, H.; Hatakeyama, T.; Sometani, M.; Okamoto, M.; Harada, S.; Okumura, H.; Yamaguchi, H. Difference in Electron Mobility at 4H–SiC/SiO_2 Interfaces with Various Crystal Faces Originating from Effective-Field-Dependent Scattering. *Appl. Phys. Lett.* **2020**, *117*, 042101. [CrossRef]
15. Yamanaka, T.; Fang, S.J.; Lin, H.-C.; Snyder, J.P.; Helms, C.R. Correlation between Inversion Layer Mobility and Surface Roughness Measured by AFM. *IEEE Electron Device Lett.* **1996**, *17*, 178–180. [CrossRef]
16. Pirovano, A.; Lacaita, A.L.; Ghidini, G.; Tallarida, G. On the Correlation between Surface Roughness and Inversion Layer Mobility in Si-MOSFETs. *IEEE Electron Device Lett.* **2000**, *21*, 34–36. [CrossRef]
17. Matsumoto, Y.; Uemura, Y. Scattering Mechanism and Low Temperature Mobility of MOS Inversion Layers. *Jpn. J. Appl. Phys.* **1974**, *13*, 367. [CrossRef]
18. Sze, S.M.; Ng, K.K. *Physics of Semiconductor Devices*, 3rd ed.; Wiley-Interscience: Hoboken, NJ, USA, 2007; ISBN 978-0-471-14323-9.
19. Sun, S.C.; Plummer, J.D. Electron Mobility in Inversion and Accumulation Layers on Thermally Oxidized Silicon Surfaces. *IEEE Trans. Electron Devices* **1980**, *27*, 1497–1508. [CrossRef]
20. Takagi, S.; Toriumi, A.; Iwase, M.; Tango, H. On the Universality of Inversion Layer Mobility in Si MOSFET's: Part I-Effects of Substrate Impurity Concentration. *IEEE Trans. Electron Devices* **1994**, *41*, 2357–2362. [CrossRef]
21. Arora, N.D.; Gildenblat, G.S. A Semi-Empirical Model of the MOSFET Inversion Layer Mobility for Low-Temperature Operation. *IEEE Trans. Electron Devices* **1987**, *34*, 89–93. [CrossRef]
22. Sabnis, A.G.; Clemens, J.T. Characterization of the Electron Mobility in the Inverted <100> Si Surface. In Proceedings of the 1979 International Electron Devices Meeting, Washington, DC, USA, 3–5 December 1979; pp. 18–21.
23. Noguchi, M.; Iwamatsu, T.; Amishiro, H.; Watanabe, H.; Miura, N.; Kita, K.; Yamakawa, S. Carrier Transport Properties in Inversion Layer of Si-Face 4H–SiC MOSFET with Nitrided Oxide. *Jpn. J. Appl. Phys.* **2019**, *58*, 031004. [CrossRef]
24. Noguchi, M.; Watanabe, T.; Watanabe, H.; Kita, K.; Miura, N. Comparative Study of Hall Effect Mobility in Inversion Layer of 4H-SiC MOSFETs With Nitrided and Phosphorus-Doped Gate Oxides. *IEEE Trans. Electron Devices* **2021**, *68*, 6321–6329. [CrossRef]
25. Das, S.; Isaacs-Smith, T.; Ahyi, A.; Kuroda, M.A.; Dhar, S. High Temperature Characteristics of Nitric Oxide Annealed P-Channel 4H-SiC Metal Oxide Semiconductor Field Effect Transistors. *J. Appl. Phys.* **2021**, *130*, 225701. [CrossRef]
26. Noguchi, M.; Iwamatsu, T.; Amishiro, H.; Watanabe, H.; Kita, K.; Miura, N. Coulomb-Limited Mobility in 4H-SiC MOS Inversion Layer as a Function of Inversion-Carrier Average Distance from MOS Interface. *Jpn. J. Appl. Phys.* **2020**, *59*, 051006. [CrossRef]
27. Uhnevionak, V.; Burenkov, A.; Strenger, C.; Ortiz, G.; Bedel-Pereira, E.; Mortet, V.; Cristiano, F.; Bauer, A.J.; Pichler, P. Comprehensive Study of the Electron Scattering Mechanisms in 4H-SiC MOSFETs. *IEEE Trans. Electron Devices* **2015**, *62*, 2562–2570. [CrossRef]
28. Potbhare, S.; Goldsman, N.; Lelis, A.; McGarrity, J.M.; McLean, F.B.; Habersat, D. A Physical Model of High Temperature 4H-SiC MOSFETs. *IEEE Trans. Electron Devices* **2008**, *55*, 2029–2040. [CrossRef]
29. Takagi, S.; Toriumi, A.; Iwase, M.; Tango, H. On the Universality of Inversion Layer Mobility in Si MOSFET's: Part II-Effects of Surface Orientation. *IEEE Trans. Electron Devices* **1994**, *41*, 2363–2368. [CrossRef]

materials

MDPI

Article

Effects of JFET Region Design and Gate Oxide Thickness on the Static and Dynamic Performance of 650 V SiC Planar Power MOSFETs

Shengnan Zhu *, Tianshi Liu, Junchong Fan, Hema Lata Rao Maddi, Marvin H. White and Anant K. Agarwal

Department of Electrical and Computer Engineering, The Ohio State University, Columbus, OH 43210, USA
* Correspondence: zhu.2670@osu.edu

Abstract: 650 V SiC planar MOSFETs with various JFET widths, JFET doping concentrations, and gate oxide thicknesses were fabricated by a commercial SiC foundry on two six-inch SiC epitaxial wafers. An orthogonal P^+ layout was used for the 650 V SiC MOSFETs to reduce the ON-resistance. The devices were packaged into open-cavity TO-247 packages for evaluation. Trade-off analysis of the static and dynamic performance of the 650 V SiC power MOSFETs was conducted. The measurement results show that a short JFET region with an enhanced JFET doping concentration reduces specific ON-resistance ($R_{on,sp}$) and lowers the gate-drain capacitance (C_{gd}). It was experimentally shown that a thinner gate oxide further reduces $R_{on,sp}$, although with a penalty in terms of increased C_{gd}. A design with 0.5 μm half JFET width, enhanced JFET doping concentration of 5.5×10^{16} cm^{-3}, and thin gate oxide produces an excellent high-frequency figure of merit (HF-FOM) among recently published studies on 650 V SiC devices.

Keywords: SiC power MOSFET; JFET width; JFET doping concentration; gate oxide thickness; orthogonal P^+ layout; gate-drain capacitance; high-frequency figure-of-merit (HF-FOM)

check for updates

Citation: Zhu, S.; Liu, T.; Fan, J.; Maddi, H.L.R.; White, M.H.; Agarwal, A.K. Effects of JFET Region Design and Gate Oxide Thickness on the Static and Dynamic Performance of 650 V SiC Planar Power MOSFETs. *Materials* **2022**, *15*, 5995. https://doi.org/10.3390/ma15175995

Academic Editors: Marilena Vivona and Mike Jennings

Received: 31 July 2022
Accepted: 22 August 2022
Published: 30 August 2022

1. Introduction

Silicon carbide (SiC) power Metal-Oxide-Semiconductor Field-Effect Transistors (MOSFETs) have been commercialized in a wide range of voltage ratings from 600 V to 1700 V. The launch of 650 V SiC MOSFETs addresses the lower voltage applications, which have traditionally been dominated by Si devices. SiC power MOSFETs outperform Si devices in low switching loss, high switching frequency, low ON-resistance (R_{on}), and high temperature operations [1–3]. Hence, designing SiC power MOSFETs with lower R_{on} and superior switching performance needs to be studied in detail.

JFET region design, including the JFET width and doping concentration, plays a crucial role in optimizing the R_{on} and switching performance of SiC MOSFETs [4]. Studies of JFET region design for 1 kV and 1.2 kV SiC MOSFETs [5,6] have demonstrated that optimizing JFET width and enhancing the doping concentration of the JFET region can reduce the JFET region resistance and lead to smaller R_{on} of SiC power MOSFETs. In addition, JFET region design affects the gate-drain capacitance (C_{gd}); C_{gd} determines the switching performance of 650 V SiC MOSFETs, primarily due to the well-known Miller effect [7]. The product of C_{gd} and R_{on} is referred to as the high-frequency figure of merit (HF-FOM) [8]. A lower HF-FOM implies better high-frequency switching performance for devices. Sung and Baliga have reported that a narrow JFET width with a high JFET doping concentration decreases C_{gd} and improves HF-FOM [9]. The gate-source and the drain-source capacitances, C_{gs} and C_{ds}, respectively, are affected by JFET width variation through the pitch of the cell, while C_{gs} and C_{ds} contribute to the switching loss of SiC power MOSFETs [10].

Gate oxide thickness plays a role in the static and dynamic performance of SiC MOSFETs. As an example, a 27 nm gate oxide was used for 650 V SiC power MOSFETs by Agarwal et al. [11,12], resulting a 1.7× better specific ON-resistance ($R_{on,sp}$) compared to MOSFETs with a 55 nm gate oxide. Under a certain operation gate voltage, a thinner gate oxide decreases $R_{on,sp}$ by reducing the channel resistance. However, a thin gate oxide increases the gate oxide capacitance (C_{ox}), and hence increases C_{gd} and C_{gs}. In addition, a thin gate oxide raises gate oxide reliability issues when sustaining high gate oxide fields [13].

In this work, the authors analyze the performance trade-offs, including threshold voltage (V_{th}), $R_{on,sp}$, breakdown voltage (BV), and parasitic capacitances for 650 V SiC MOSFETs with different JFET widths, JFET doping concentrations, and gate oxide thicknesses. The 650 V SiC power MOSFETs were fabricated on two six-inch SiC epitaxial wafers by a commercial foundry. The design details and fabrication information are presented in Section 2. The preliminary wafer-level characterizations have been published in [14]. The fabricated devices were packaged for static and dynamic measurements. The experimental methods are explained in Section 3. In Section 4, the experimental results are presented and discussed. Section 5 provides further analysis of the performance trade-offs for the 650 V SiC MOSFETs.

2. Device Design and Fabrication

The layout design of a 650 V SiC MOSFET is shown in Figure 1a. The layout is in a stripe pattern, with square P^+ regions located periodically in the center of the P-well stripe. The orthogonal P^+ layout reduces the R_{on} of the MOSFETs by reducing the cell pitch compared to the traditional linear striped P^+ layout. The cross-section along the A-A′ cutline is shown in Figure 1b. The half-cell pitch consists of P^+ width (1 μm), N^+ source width (1.1 μm), channel length (0.5 μm), and half JFET width ($\frac{1}{2}W_{JFET}$). The spacing between the source contact and polysilicon gate is 0.7 μm. The ohmic contact width is 1 μm. The cross-section along B-B′ (Figure 1c) shows the layout with only N^+ source. The extended N^+ source replaces the P^+ in the A-A′ half-cell pitch and produces a total N^+ source width of 2.1 μm. Four devices with different half-JFET widths were designed ($\frac{1}{2}W_{JFET}$ = 0.4, 0.5, 0.6, and 0.75 μm).

Twenty-two P^+ guard rings were used as the edge termination for all layouts. Each guard ring had a width of 2 μm. A cross-sectional view of the edge termination is shown in Figure 1d. The edge termination can be divided into four sections. The spacing for each section is illustrated in Figure 1d; spacing was identical in each section. The total length of the edge termination was 77.6 μm.

Different JFET doping concentrations (N_{JFET}) and gate oxide thicknesses (t_{ox}) were utilized during the fabrication of the devices. The devices were fabricated on two six-inch 4H-SiC wafers (wafer 1 and wafer 2) with n-type epitaxial layers on N^+ substrates. The substrates were thinned to reduce the resistance. The epitaxial layer was doped with nitrogen with a doping concentration of 2×10^{16} cm^{-3}. Ion implantation of nitrogen was used to form the JFET region and N^+ source; $N_{JFET} = 4 \times 10^{16}$ cm^{-3} and $N_{JFET} = 5.5 \times 10^{16}$ cm^{-3} were used for wafers 1 and 2, respectively. Aluminum ions were implanted to form the P-well and P^+ region. The gate oxide was grown after the implantation and activation annealing processes. The gate oxide thicknesses on wafers 1 and 2 are represented as t_{ox1} (36∼44 nm) and t_{ox2} (32∼38 nm), respectively; t_{ox2} is 12.5% less than t_{ox1}. Details of the gate oxide thicknesses have been discussed previously in [14]. Self-alignment technology was utilized to form the MOS channel. Fabrication was completed following the standard process flow of commercial SiC MOSFETs.

The design parameters and experimental results for all devices are summarized in Table 1 (Section 5). Figure 2a shows a cross-sectional SEM image (along BB′ in Figure 1c) of the fabricated 650 V SiC MOSFET ($\frac{1}{2}W_{JFET}$ = 0.6 μm) on wafer 1. Due to the lateral straggle of Aluminum implantation in the P-well, the narrowest portion of the JFET region is reduced by 0.2 μm on each side.

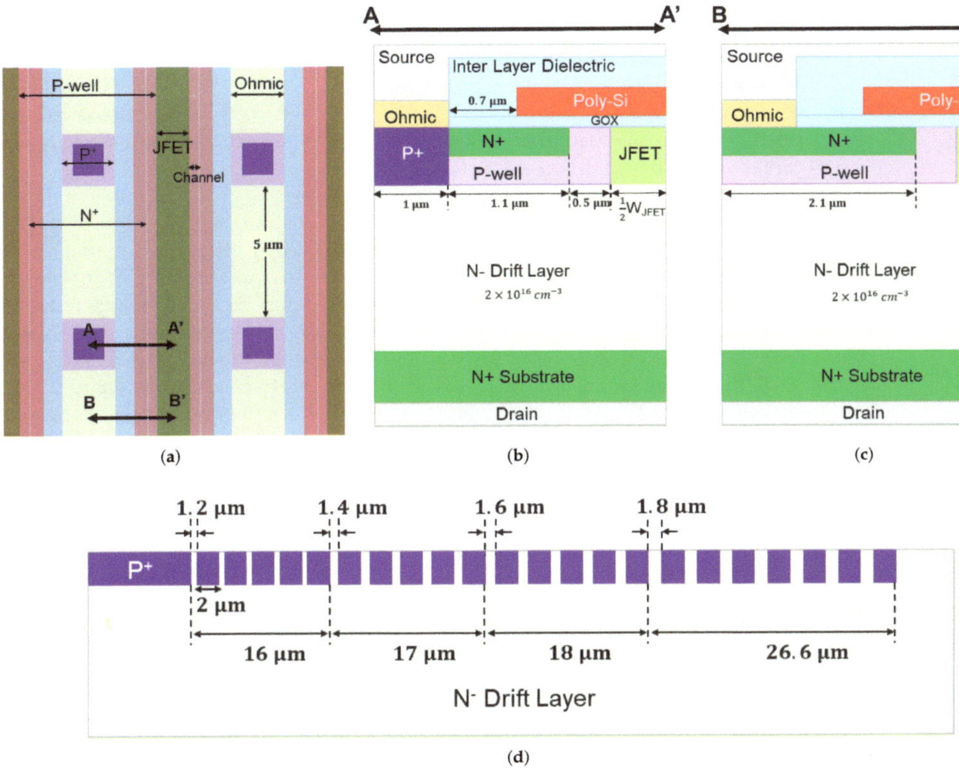

Figure 1. (**a**) Layout design of a SiC power MOSFET with P^+ located periodically in the center of P-well stripe; (**b**) A-A′ cross-sectional view showing both P^+ and N^+; (**c**) B-B′ cross-sectional view showing extended N^+ source; (**d**) cross-sectional view of the edge termination of the fabricated 650 V SiC power MOSFETs.

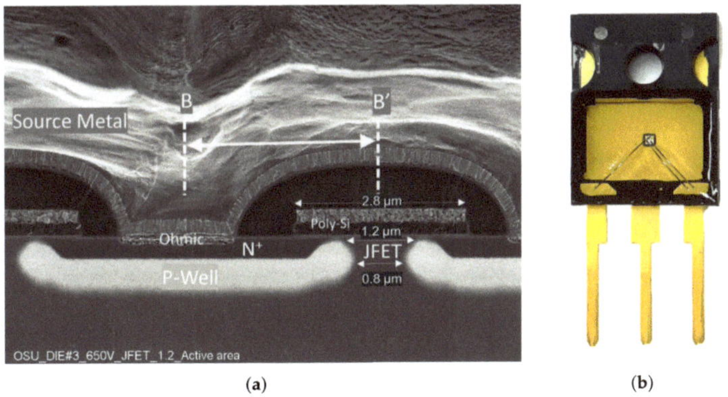

Figure 2. (**a**) Cross-sectional SEM image along B-B′ of the fabricated 650 V SiC power MOSFETs with $\frac{1}{2}W_{JFET}$ = 0.6 µm and (**b**) 650 V SiC power MOSFET in a open-cavity TO-247 package.

3. Experimental Methods

3.1. Device Packaging

The fabricated MOSFETs were diced and packaged into open cavity TO-247 packages, as shown in Figure 2b. A single 5-mil aluminum wire bond was used for the gate terminal, while two-wire bonds were attached on the the source area to decrease the parasitic resistance. Silicone dielectric gel was used to fill the cavity to protect the bare die. Five copies of each layout design on wafers 1 and 2 were packaged.

3.2. Device Characterization

The static performance of the MOSFETs, including the transfer, output, and blocking characteristics, were measured with a Keysight B1506A semiconductor parameter analyzer. We extracted V_{th} at a drain current of 1 mA from the transfer characteristics tested under a drain bias of 100 mV. The output characteristics were measured under a gate bias of 20 V, with the drain voltage swept from 0 to 2 V. We obtained the R_{on} of the device under test (DUT) at a drain bias of 1.5 V; BV was obtained from the blocking I-V characteristics at a current of 100 µA, while C_{gd}, C_{gs}, and C_{ds} were measured up to a drain bias of 400 V at a frequency of 100 kHz using a Keysight B1505A semiconductor parameter analyzer.

4. Device Characteristics and Discussion

The measured device characteristics for the packaged 650 V SiC MOSFETs with different designs are illustrated and compared in this section.

4.1. Threshold Voltage

The transfer characteristics for the devices with different $\frac{1}{2}W_{JFET}$ on wafer 1 are plotted in Figure 3a. Typical transfer curves of SiC MOSFETs were obtained from all the DUTs. The average V_{th} from the five copies of each design is plotted in Figure 3b. Minimal V_{th} variation was observed for wafers 1 and 2 when increasing $\frac{1}{2}W_{JFET}$.

The V_{th} of the MOSFETs on wafer 2 is ~0.5 V smaller than wafer 1, as shown in Figure 3b. The thinner gate oxide contributes to the V_{th} reduction; here, V_{th} is defined as [8]:

$$V_{\mathrm{th}} = \Phi_{\mathrm{MS}} + \frac{\sqrt{4\varepsilon_{\mathrm{SiC}} kTN_{\mathrm{A}}\mathrm{In}(N_{\mathrm{A}}/n_{\mathrm{i}})} - Q_{\mathrm{ox}}}{C_{\mathrm{ox}}} + \frac{2kT}{q}\mathrm{In}(\frac{N_{\mathrm{A}}}{n_{\mathrm{i}}}), \quad (1)$$

where Φ_{MS} is the metal–semiconductor work function difference, ε_{ox} is the permittivity of SiC, k and T are the Boltzmann constant and temperature, respectively, n_i is the intrinsic carrier concentration of SiC, q is the electric charge, Q_{ox} is the total effective charge in the oxide (the sum of the fixed and interface charges), and N_A is the net p-type doping concentration at the channel region; C_{ox} is the gate oxide capacitance, which is given as

$$C_{\mathrm{ox}} = \frac{\varepsilon_{\mathrm{ox}}}{t_{\mathrm{ox}}} \quad (2)$$

where ε_{ox} is the permittivity of oxide. Comparing wafer 2 to wafer 1, higher N_{JFET} reduces N_A by the effect of the counter doping at the surface. Additionally, the thinner gate oxide of wafer 2 increases C_{ox}. According to (1), the reduced N_A at the surface and increased C_{ox} lead to smaller V_{th} of the MOSFETs on wafer 2.

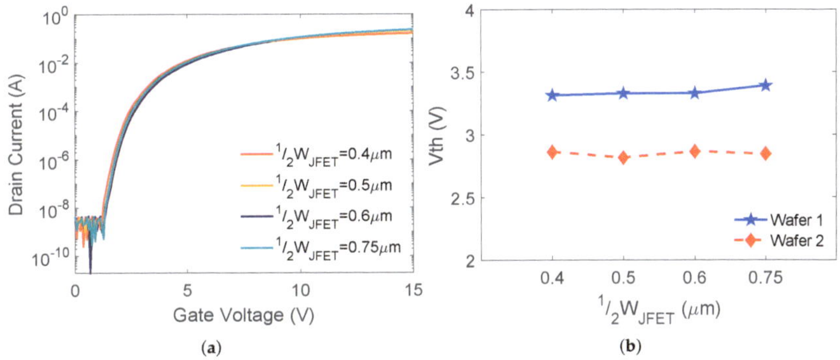

Figure 3. (**a**) Measured transfer characteristics of the packaged 650 V SiC MOSFETs on wafer 1 and (**b**) V_{th} variation as a function of $\frac{1}{2}W_{JFET}$ for MOSFETs on wafers 1 and 2.

4.2. Specific ON-Resistance

Figure 4a shows the output characteristics at a gate bias of 20 V for devices on wafer 1. Drain current increases with a wider JFET region. Figure 4b plots $R_{on,sp}$ versus $\frac{1}{2}W_{JFET}$ variation. For both wafers 1 and 2, $R_{on,sp}$ is reduced when increasing $\frac{1}{2}W_{JFET}$ because a larger $\frac{1}{2}W_{JFET}$ provides lower JFET region resistance [14]. With the same $\frac{1}{2}W_{JFET}$, $R_{on,sp}$ reduction from wafer 1 to wafer 2 is contributed by thinner gate oxide and higher N_{JFET}. A considerable (1.6×) $R_{on,sp}$ reduction is observed when $\frac{1}{2}W_{JFET}$ rises from 0.4 μm to 0.5 μm on wafer 1, while the tendency is weaker for wafer 2. These results indicate that thinner gate oxide and higher N_{JFET} make $R_{on,sp}$ less susceptible to $\frac{1}{2}W_{JFET}$ variation.

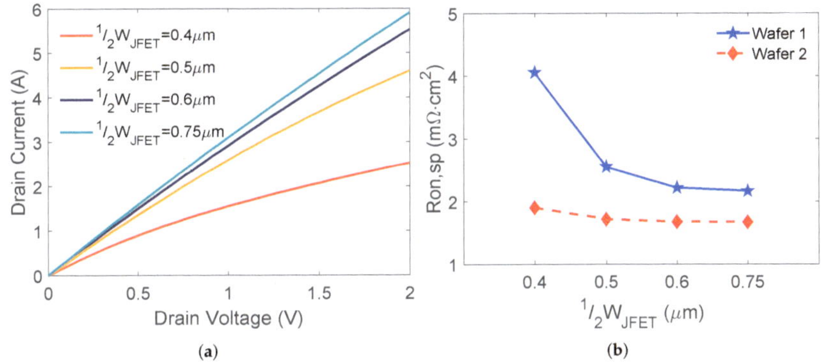

Figure 4. (**a**) Measured output characteristics of the SiC MOSFETs on wafer 1 and (**b**) $R_{on,sp}$ variation as a function of $\frac{1}{2}W_{JFET}$ for MOSFETs on wafers 1 and 2.

4.3. Breakdown Voltage

The blocking characteristics for MOSFETs on wafer 1 are shown in Figure 5a. All DUTs maintain low leakage currents (∼100 pA) under drain voltage up to 550 V. The drain to source breakdown of a planar SiC MOSFET is triggered by avalanche breakdown, and both N_{JFET} and $\frac{1}{2}W_{JFET}$ have little effect on the BV determined by avalanche breakdown [5]. Our experimental results (Figure 5b) show that the BV of 650 V SiC MOSFETs is minimally changed by $\frac{1}{2}W_{JFET}$ variation.

A maximum BV of about 780 V is achieved for devices on wafer 1. The BV for MOSFETs on wafer 2 is ∼640 V. The 18% BV drop from wafer 1 to wafer 2 is mainly caused by the difference in drift layer doping. The drift layer doping concentrations can be extracted from the C-V measurement of MOS capacitors on both wafers [15]. The extracted drift layer doping concentrations are 1.8×10^{16} cm^{-3} and 2.1×10^{16} cm^{-3} for wafers 1 and 2, respectively. This difference explains the reduction of BV on wafer 2.

Although BV does not change with $\frac{1}{2}W_{JFET}$ variation, a smaller $\frac{1}{2}W_{JFET}$ improves the gate oxide reliability of the MOSFETs by better shielding the gate oxide on the top of the JFET region from high oxide fields under the blocking condition [14,16]. These high oxide fields may cause high gate leakage currents, degrade the gate oxide, and reduce the oxide lifetime [13,17], and can lead to failures during High-Temperature Reverse Bias (HTRB) testing.

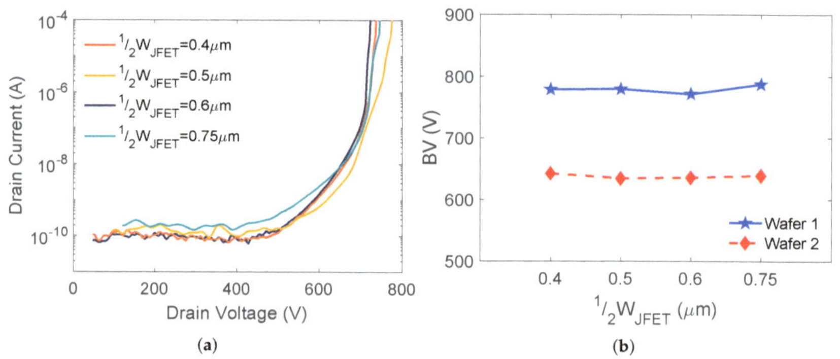

Figure 5. (**a**) Measured blocking characteristics of the SiC MOSFETs on wafer 1 and (**b**) Maximum BV as a function of $\frac{1}{2}W_{JFET}$ for MOSFETs on wafers 1 and 2.

4.4. Device Capacitances

The device capacitances as a function of the applied drain bias for 650 V MOSFETs on wafer 1 are shown in Figure 6a. As expected, the measured C_{gd} and C_{ds} are nonlinear functions of the drain bias, while C_{gs} stays relatively constant with increasing drain bias. The extracted C_{gd}, C_{ds}, and C_{gs} as function of $\frac{1}{2}W_{JFET}$ for the MOSFETs on wafers 1 and 2 are shown in Figure 6b–d, respectively.

When extending $\frac{1}{2}W_{JFET}$, C_{gd} increases. For a planar SiC MOSFET, C_{gd} is formed by the overlap between the gate and drain electrodes. A complete cell cross-section in Figure 7, illustrating the various device capacitance components. Here, C_{gd} is composed of C_{ox} and depletion region capacitance ($C_{SiC,MOS}$) under the gate oxide; C_{gd} is defined as follows:

$$C_{gd} = \frac{W_{JFET}}{W_{cell}}\left(\frac{C_{ox} \cdot C_{SiC,MOS}}{C_{ox} + C_{SiC,MOS}}\right) \cdot A_{active}. \quad (3)$$

Equation (3) is based on [8], where W_{cell} refers to the cell pitch, A_{active} represents the active area of the device, C_{ox} stays constant for the devices with the same t_{ox}, and $C_{SiC,MOS}$ is determined by the depletion layer thickness under the gate oxide, which does not change for devices with the same N_{JFET} and which sustain a specific drain bias. According to (3), C_{gd} increases when increasing W_{JFET}, which agrees with the measured results for both wafers 1 and 2 in Figure 6b.

Comparing wafer 2 to wafer 1, t_{ox} drops by 12.5%. Correspondingly, C_{ox} increases by 14.3% and leads to C_{gd} increasing. The enhanced N_{JFET} of wafer 2 affects $C_{SiC,MOS}$ by changing the thickness and the width of the depletion layer [7,18]. It is challenging to identify the change of $C_{SiC,MOS}$ quantitatively, as the depletion layer varies with the gate-drain bias, p-well potential, and doping concentration of the JFET and drift layer [19]. The results in [9] demonstrate that a higher N_{JFET} leads to a higher C_{gd}. Thus, the overall outcome from lowering t_{ox} and increasing N_{JFET} is the increase of C_{gd}. The measured C_{gd} for MOSFETs on wafer 2 is about 1.4× higher than those of wafer 1 under a given $\frac{1}{2}W_{JFET}$, as shown in Figure 6b.

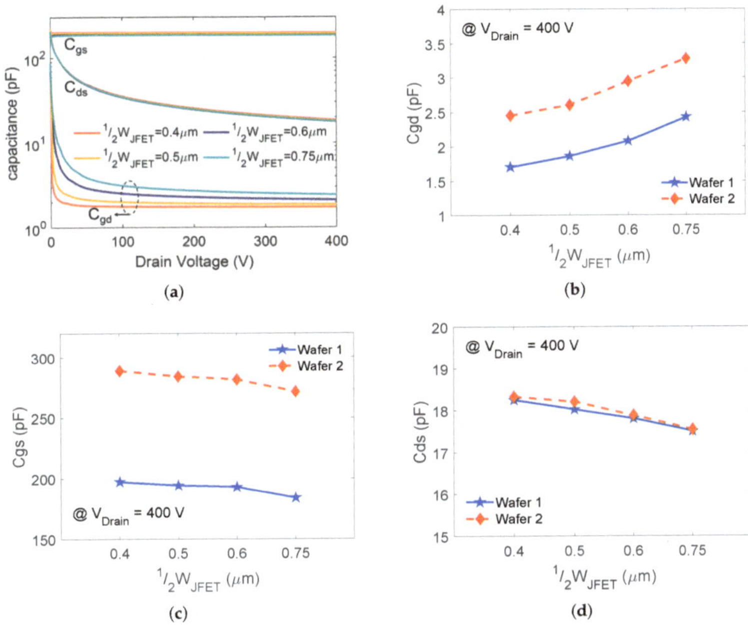

Figure 6. (**a**) Measured device capacitances vs. drain voltage at 100 kHz of the SiC MOSFETs on wafer 1, (**b**) C_{gd}, (**c**) C_{ds}, and (**d**) C_{gs} variation as a function of $\frac{1}{2}W_{JFET}$ for MOSFETs on wafers 1 and 2.

Note that C_{gs} consists of the overlap capacitance of the gate electrode with source plus channel region and the parallel capacitance across the gate and source metallization (C_{ILD}) [20]; C_{gs} in the active area of a 650 V SiC MOSFET is addressed as

$$C_{gs} = (\frac{2W_{GS}}{W_{cell}}C_{ox} + \frac{2W_{GS} + W_{JFET}}{W_{cell}}C_{ILD}) \cdot A_{active}, \tag{4}$$

where W_{GS} is the total length of the overlap between gate and N^+ source and the channel region and C_{ILD} is the inter-layer dielectric capacitance, which stays constant for all the devices due to the same fabrication process being used for wafers 1 and 2.

Among the designs on the same wafer, increasing W_{JFET} reduces the coefficients of C_{ox} and C_{ILD} in (4) and leads to the increase of C_{gs}. The measured C_{gs} verifies the variation for both wafer 1 and wafer 2 in Figure 6c. For a specific W_{JFET}, a thinner gate oxide increases C_{ox}, and hence result in a higher C_{gs} according to (4). This explains the higher C_{gs} measured on wafer 2 compared to C_{gs} on wafer 1.

As C_{ds} is driven by the depletion layer formation at the P-well and drift region interface, the total C_{ds} in the active area of a 650 V SiC MOSFET is expressed as

$$C_{ds} = \frac{W_{cell} - W_{JFET}}{W_{cell}}C_J \cdot A_{active}, \tag{5}$$

where C_J is the junction capacitance per unit area, which is determined by the depletion layer thickness. All the DUTs in this work have similar doping concentration of the epi-layer. The bottom of the P-well is heavily doped, meaning that the depletion thickness in the p-well region can be neglected. Thus, under a particular drain bias, the depletion layer thickness stays almost the same for all DUTs, which results in similar C_J. According to (5), increasing W_{JFET} reduces C_{ds}, corresponding to the measured results in Figure 6d for both wafer 1 and wafer 2. In addition, the measured C_{ds} under a certain $\frac{1}{2}W_{JFET}$ is almost the same for the MOSFETs on both wafers, which is due to the fact that t_{ox} and N_{JFET} are not involved in (5).

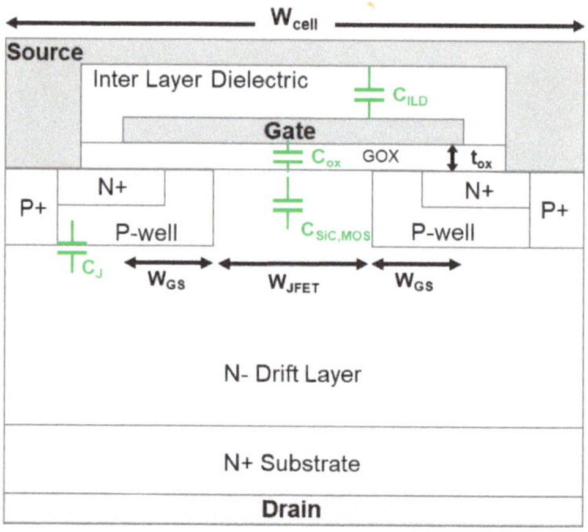

Figure 7. Device capacitance components.

5. Trade-Offs

Table 1 summarizes the design information and experimental results for the 650 V SiC MOSFETs. HF-FOM is included to evaluate the performance of the devices.

The variation of $\frac{1}{2}W_{JFET}$ influences $R_{on,sp}$ and the device capacitance. When reducing the $\frac{1}{2}W_{JFET}$ from 0.75 μm to 0.4 μm, (1) $R_{on,sp}$ increases by 1.9× for wafer 1 and 1.1× for wafer 2; (2) C_{gd} decreases by 1.4× for wafer 1 and 1.3× for wafer 2; and (3) less than 7% and 4% increase are identified for C_{gs} and C_{ds}, respectively.

Comparing the performance of the MOSFETs on wafer 2 to those on wafer 1, higher N_{JFET} and thinner gate oxide have the following benefits: (1) $R_{on,sp}$ is further reduced and the variation of $R_{on,sp}$ caused by variation in $\frac{1}{2}W_{JFET}$ is mitigated; (2) V_{th} is reduced by about 10%; and (3) a low HF-FOM of 699 mΩ·pF is obtained at $\frac{1}{2}W_{JFET}$ of 0.5 μm. The trade-offs are that C_{gs} and C_{gd} are increased and the oxide field on the top of the JFET region may rise; C_{ds} is not affected. BV should not be affected either, assuming that the drift layer doping and thickness remain the same.

Combining the above analysis, a narrower JFET region with a thinner gate oxide and enhanced N_{JFET} produce optimized designs for 650 V SiC MOSFETs. A small W_{JFET} reduces C_{gd}. The increased $R_{on,sp}$ thanks to smaller W_{JFET} can be compensated for by thinner gate oxide and higher N_{JFET}. A narrow JFET region helps to shield the gate oxide on the top of JFET region from high oxide fields that may be induced by the thin gate oxide and high N_{JFET}.

Table 1. Summary of design information and experimental results.

	Design Information					Experimental Results													
	$\frac{1}{2}W_{JFET}$ [µm]	t_{ox} [nm]	N_{JFET} [cm^{-3}]	Cell Pitch [µm]	Active Area [mm^2]	$R_{on,sp}$ [$m\Omega\cdot cm^2$]	$R_{on,sp}$ Std.	V_{th} [V]	V_{th} Std.	BV [V]	BV Std.	C_{gs} [pF]	C_{gs} Std.	C_{ds} [pF]	C_{ds} Std.	C_{gd} [pF]	C_{gd} Std.	HF-FOM ($C_{gd} \times R_{on}$) [$m\Omega\cdot pF$]
Wafer 1	0.4	t_{ox1} *	4×10^{16}	6	0.64	4.06	0.77	3.3	0.08	780	20.5	197	4.0	18.3	0.2	1.7	0.05	1078
	0.5			6.2	0.64	2.55	0.07	3.3	0.09	780	19.1	194	2.6	18.0	0.3	1.9	0.05	757
	0.6			6.4	0.64	2.22	0.03	3.3	0.08	772	19.9	193	2.9	17.8	0.2	2.1	0.03	728
	0.75			6.7	0.64	2.16	0.04	3.4	0.10	788	21.7	184	3.0	17.5	0.3	2.4	0.03	810
Wafer 2	0.4	t_{ox2} *	5.5×10^{16}	6	0.64	1.90	0.08	2.9	0.11	643	31.6	289	1.5	18.3	0.3	2.5	0.05	742
	0.5			6.2	0.64	1.72	0.03	2.8	0.10	635	36.8	284	2.5	18.2	0.3	2.6	0.05	699
	0.6			6.4	0.64	1.68	0.03	2.9	0.09	637	30.7	282	1.8	17.9	0.4	3.0	0.05	788
	0.75			6.7	0.64	1.67	0.03	2.9	0.11	640	45.0	272	2.0	17.6	0.3	3.3	0.03	861

* t_{ox2} is 12.5% less than t_{ox1}.

6. Conclusions

In this paper, 650 V SiC MOSFETs were designed, fabricated, packaged, and characterized. The on-state and dynamic performance trade-offs due to the JFET region and gate oxide thickness design were then analyzed. Our experimental results show that a narrow JFET width and enhanced JFET doping concentration lead to low $R_{on,sp}$, low C_{gd}, low HF-FOM, and better gate oxide reliability without degrading the V_{th} and BV. The increases in C_{gs} and C_{ds} with reduction in JFET width are relatively small in comparison with the reduction of C_{gd}. In addition, we have shown that $R_{on,sp}$ can be further reduced with a thinner gate oxide, although this incurs a penalty in terms of increased C_{gd}.

Author Contributions: Conceptualization, S.Z., T.L. and A.K.A.; methodology, S.Z., T.L. and A.K.A.; investigation, S.Z. and J.F.; resources, H.L.R.M.; data curation, S.Z.; writing—original draft preparation, S.Z.; writing—review and editing, T.L., M.H.W. and A.K.A. All authors have read and agreed to the published version of the manuscript.

Funding: This research was funded in part by a Block Gift Grant from II-VI Foundation and in part by the Ford Motor Company under the Ford Alliance 2019 Project to The Ohio State University.

Institutional Review Board Statement: Not applicable.

Informed Consent Statement: Not applicable.

Data Availability Statement: Not applicable.

Acknowledgments: We acknowledge Diang Xing for his helpful discussion.

Conflicts of Interest: The authors declare no conflict of interest.

Abbreviations

The following abbreviations are used in this manuscript:

SiC	Sillicon Carbide
MOSFET	Metal–Oxide Semiconductor Field-Effect Transistor
JFET	Junction Field Effect Transistor
HF-FOM	High-Frequency Figure Of Merit
DUT	Device Under Test

References

1. Kimoto, T. Material science and device physics in SiC technology for high-voltage power devices. *Jpn. J. Appl. Phys.* **2015**, *54*, 040103. [CrossRef]
2. Zhang, L.; Yuan, X.; Wu, X.; Shi, C.; Zhang, J.; Zhang, Y. Performance evaluation of high-power SiC MOSFET modules in comparison to Si IGBT modules. *IEEE Trans. Power Electron.* **2018**, *34*, 1181–1196. [CrossRef]
3. Maddi, H.L.R.; Yu, S.; Zhu, S.; Liu, T.; Shi, L.; Kang, M.; Xing, D.; Nayak, S.; White, M.H.; Agarwal, A.K. The Road to a Robust and Affordable SiC Power MOSFET Technology. *Energies* **2021**, *14*, 8283. [CrossRef]
4. Vathulya, V.R.; Shang, H.; White, M.H. A novel 6H-SiC power DMOSFET with implanted p-well spacer. *IEEE Electron Device Lett.* **1999**, *20*, 354–356. [CrossRef]
5. Saha, A.; Cooper, J.A. A 1-kV 4H-SiC power DMOSFET optimized for low on-resistance. *IEEE Trans. Electron Devices* **2007**, *54*, 2786–2791. [CrossRef]
6. Kim, D.; Jang, S.Y.; Morgan, A.J.; Sung, W. An inclusive structural analysis on the design of 1.2 kV 4H-SiC planar MOSFETs. *IEEE J. Electron Devices Soc.* **2021**, *9*, 804–812. [CrossRef]
7. Wu, L.; Xiao, L.; Zhao, J.; Chen, G. Physical analysis and modeling of the nonlinear miller capacitance for SiC MOSFET. In Proceedings of the IECON 2017-43rd Annual Conference of the IEEE Industrial Electronics Society, Beijing, China, 29 October–1 November 2017; pp. 1411–1416.
8. Baliga, B.J. *Fundamentals of Power Semiconductor Devices*; Springer Science & Business Media: Berlin/Heidelberg, Germany, 2010.
9. Sung, W.; Han, K.; Baliga, B. Optimization of the JFET region of 1.2 kV SiC MOSFETs for improved high frequency figure of merit (HF-FOM). In Proceedings of the 2017 IEEE 5th Workshop on Wide Bandgap Power Devices and Applications (WiPDA), Albuquerque, NM, USA, 30 October–1 November 2017; pp. 238–241.
10. Li, X.; Zhang, L.; Guo, S.; Lei, Y.; Huang, A.Q.; Zhang, B. Understanding switching losses in SiC MOSFET: Toward lossless switching. In Proceedings of the 2015 IEEE 3rd Workshop on Wide Bandgap Power Devices and Applications (WiPDA), Blacksburg, VA, USA, 2–4 November 2015; pp. 257–262.

11. Agarwal, A.; Kanale, A.; Baliga, B.J. Advanced 650 V SiC Power MOSFETs With 10 V Gate Drive Compatible With Si Superjunction Devices. *IEEE Trans. Power Electron.* **2020**, *36*, 3335–3345. [CrossRef]
12. Agarwal, A.; Kanale, A.; Han, K.; Baliga, B.J. Switching and short-circuit performance of 27 nm gate oxide, 650 V SiC planar-gate MOSFETs with 10 to 15 V gate drive voltage. In Proceedings of the 2020 32nd International Symposium on Power Semiconductor Devices and ICs (ISPSD), Vienna, Austria, 13–18 September 2020; pp. 250–253.
13. Zhu, S.; Liu, T.; White, M.H.; Agarwal, A.K.; Salemi, A.; Sheridan, D. Investigation of gate leakage current behavior for commercial 1.2 kv 4h-sic power mosfets. In Proceedings of the 2021 IEEE International Reliability Physics Symposium (IRPS), Monterey, CA, USA, 21–25 March 2021; pp. 1–7.
14. Liu, T.; Zhu, S.; Salemi, A.; Sheridan, D.; White, M.H.; Agarwal, A.K. JFET Region Design Trade-Offs of 650 V 4H-SiC Planar Power MOSFETs. *Solid State Electron. Lett.* **2021**, *3*, 53–58. [CrossRef]
15. Schroder, D.K. *Semiconductor Material and Device Characterization*; John Wiley & Sons: Hoboken, NJ, USA, 2015.
16. Zhu, Z.; Ren, N.; Xu, H.; Liu, L.; Sheng, K. Avalanche Reliability of Planar-gate SiC MOSFET with Varied JFET Region Width and Its Balance with Characteristic Performance. In Proceedings of the 2021 33rd International Symposium on Power Semiconductor Devices and ICs (ISPSD), Nagoya, Japan, 30 May–3 June 2021; pp. 231–234.
17. Liu, T.; Zhu, S.; White, M.H.; Salemi, A.; Sheridan, D.; Agarwal, A.K. Time-dependent dielectric breakdown of commercial 1.2 kv 4h-sic power mosfets. *IEEE J. Electron Devices Soc.* **2021**, *9*, 633–639. [CrossRef]
18. Li, H.; Jiang, Y.; Qiu, Z.; Wang, Y.; Ding, Y. A predictive algorithm for crosstalk peaks of SiC MOSFET by considering the nonlinearity of gate-drain capacitance. *IEEE Trans. Power Electron.* **2020**, *36*, 2823–2834. [CrossRef]
19. Shintani, M.; Nakamura, Y.; Hiromoto, M.; Hikihara, T.; Sato, T. Measurement and modeling of gate–drain capacitance of silicon carbide vertical double-diffused MOSFET. *Jpn. J. Appl. Phys.* **2017**, *56*, 04CR07. [CrossRef]
20. Farhadi, M.; Yang, F.; Pu, S.; Vankayalapati, B.T.; Akin, B. Temperature-independent gate-oxide degradation monitoring of SiC MOSFETs based on junction capacitances. *IEEE Trans. Power Electron.* **2021**, *36*, 8308–8324. [CrossRef]

materials

MDPI

Article

Thorough Wide-Temperature-Range Analysis of Pt/SiC and Cr/SiC Schottky Contact Non-Uniformity

Razvan Pascu [1,2], Gheorghe Pristavu [2,*], Dan-Theodor Oneata [2], Gheorghe Brezeanu [2,*], Cosmin Romanitan [1], Nikolay Djourelov [3], Andrei Enache [2], Florin Draghici [2], Andrei Mario Ivan [4] and Emilian Ceuca [5]

1 National Institute for Research and Development in Microtechnologies—IMT Bucharest, 126A, Erou Iancu Nicolae Street, 077190 Bucharest, Romania; razvan.pascu@imt.ro (R.P.); cosmin.romanitan@imt.ro (C.R.)

2 Faculty of Electronics, Telecommunications and Information Technology, National University of Science and Technology Politehnica Bucharest, 060042 Bucharest, Romania; dan_theodor.oneata@upb.ro (D.-T.O.); andrei.enache1512@upb.ro (A.E.); florin.draghici@upb.ro (F.D.)

3 Extreme Light Infrastructure-Nuclear Physics (ELI-NP), Horia Hulubei National R&D Institute for Physics and Nuclear Engineering (IFIN-HH), 077125 Magurele, Romania; nikolay.djourelov@eli-np.ro

4 Faculty of Industrial Engineering and Robotics, National University of Science and Technology Politehnica Bucharest, 060042 Bucharest, Romania; andrei_mario.ivan@upb.ro

5 Department of Informatics, Mathematics and Electronics, Faculty of Exact Sciences and Engineering, University "1 Decembrie 1918" of Alba Iulia, No. 5 Gabriel Bethlen Street, 510009 Alba Iulia, Romania; emilian.ceuca@uab.ro

* Correspondence: gheorghe.pristavu@upb.ro (G.P.); gheorghe.brezeanu@upb.ro (G.B.)

Abstract: This paper evaluates the non-uniformity degree of platinum and chromium Schottky contacts on silicon carbide. The forward characteristics of experimental samples were acquired in a wide, 60–500 K, temperature range. Microstructural and conventional electrical characterizations were performed, revealing the presence of inhomogeneities on the contact surface. The main parameters were extracted using inhomogeneity models of varying complexity levels. Their relevance is discussed with respect to the models' applicable, limited, temperature ranges. Finally, complete forward curve fitting was achieved using *p-diode* modeling, evincing that each type of contact behaves as four parallel-connected ideal diodes. Since these parallel diodes have varying influences on the overall device current with temperature and bias, operable domains can be identified where the samples behave suitably.

Keywords: Schottky contact; silicon carbide; *p-diode* model; non-uniformity

Citation: Pascu, R.; Pristavu, G.; Oneata, D.-T.; Brezeanu, G.; Romanitan, C.; Djourelov, N.; Enache, A.; Draghici, F.; Ivan, A.M.; Ceuca, E. Thorough Wide-Temperature-Range Analysis of Pt/SiC and Cr/SiC Schottky Contact Non-Uniformity. *Materials* **2024**, *17*, 400. https://doi.org/10.3390/ma17020400

Academic Editors: Marilena Vivona and Mike Jennings

Received: 20 December 2023
Revised: 4 January 2024
Accepted: 9 January 2024
Published: 13 January 2024

1. Introduction

Reliable operation in harsh environments and over a wide range of temperatures is a fundamental requirement in many industrial applications [1,2], including cement manufacturing [3], drilling [4], geothermal systems [5], aerospace [6], etc. The working conditions in these applications often include strong vibrations, corrosion, radiation, and elevated heat levels, as well as a significant number of thermal cycles for the involved devices [3].

For such hostile environments, silicon carbide (SiC)-based sensors have emerged as a promising solution [3], especially for applications involving gas concentration measurement [7,8] or temperature sensing [3,9–12]. Due to the wide bandgap (3.24 eV for 4H-SiC) and low intrinsic carrier concentration, SiC devices can operate at temperatures far above the limits for conventional semiconductors [3]. Furthermore, the mechanical robustness, radiation hardness, chemical inertness, and high thermal conductivity of SiC allow for the operation of these devices under harsh conditions [13].

The simplest semiconductor device that can be fabricated is the Schottky barrier diode (SBD), as the process involves a metal deposition followed by annealing in order to obtain a rectifying contact [14]. Consequently, an SBD is also the most cost-effective and technologically mature device fabricated using SiC [3]. Reported applications range from high-voltage, high-power circuits, such as traction inverters [15], to sensors destined for harsh environmental conditions [3,7].

Over time, different metals have been used for the Schottky contact on SiC substrates, including nickel [3], titanium [16], platinum [17,18] and, less frequently, chromium [19,20]. While Pt is suitable for applications in a wide temperature range, Cr is also an attractive proposition as a Schottky metal because it forms both silicides and carbides with SiC (very stable compounds), resulting in good contact adhesion and mechanical properties [19].

Irrespective of the utilized metal, and even after annealing, contact inhomogeneity still appears, with an observable effect on the Schottky barrier height (SBH) [14,21,22]. Consequently, evaluating the performance of an SBD is still challenging, despite there being more than fifty years of investigations [23]. Therefore, in order to fully understand the behavior of these inhomogeneous contacts, comprehensive characterization needs to be carried out over large temperature spans. The resulting characteristics must be parameterized using specialized models, which need to accurately explain the electrical behavior across the entire range. Modeling accuracy is especially critical for sensing applications, where the SBD model must allow for the precise determination of the sensed quantity (temperature, gas concentration, etc.) based on the measured electrical quantity (voltage or current); thus, the model needs to account for the effects of contact inhomogeneity.

In this paper, the fabrication and electrical characterization of Pt/4H-SiC and Cr/4H-SiC Schottky barrier diodes are presented. The characterization was carried out over a broad temperature domain (60–500 K). The resulting forward curves were comprehensively parameterized using state-of-the-art methods, confirming the presence of contact inhomogeneity.

Afterwards, complete curve fitting was carried out using our recently proposed *p-diode* technique [14,24], which modeled the inhomogeneous SBD as a minimal number of parallel diodes. The resulting excellent fitting accuracy demonstrates that the *p-diode* model can fully explain the forward behavior of the sample over the entire temperature and bias intervals.

2. Materials and Methods

In this work, two metals with major differences in their work function (WF) values were used to fabricate Schottky barrier diodes on nitrogen-doped 4H-SiC wafers with an 8 μm epitaxial layer. The SiC substrate was heavily doped, and the epitaxial layer had a doping concentration of around 10^{16} cm^{-3}. After the standard RCA chemical cleaning was performed, two different types of SiO_2 layers were deposited by the low-pressure chemical vapor deposition (LPCVD) method in order to obtain a ramp profile as a termination for the Schottky contact. This technological process is described in detail in refs. [3,25]. The Schottky contact had a circular configuration with a diameter of 400 μm [25], as defined by standard photolithography techniques (lift-off for Pt and wet etching for Cr contacts) in the deposited field oxide (LPCVD). Due to the very high annealing temperature constraint, the ohmic contact was firstly defined. Thus, 100 nm of Ni was deposited by sputtering on the wafer backside, which was followed by rapid thermal annealing at 1050 °C for 3 min in an Ar atmosphere. An X-ray diffraction analysis evinced a nickel silicide compound with multiple diffraction peaks, which were assigned unambiguously to the Ni_2Si phase [25]. For the Schottky contact, the two metals (Pt with WF $\cong$ 5.7 eV and Cr with WF $\cong$ 4.5 eV) with a thickness of 100 nm were deposited into the circular windows using an e-beam evaporation system. Annealing in the same conditions (600 °C for 3 min in an Ar atmosphere) was performed for the diodes from both batches, which were henceforth named *Pt/4H-SiC* and *Cr/4H-SiC*. The pad contacts and backside metallization of the final structures were achieved by evaporation of a metallic stack consisting of Ti (60 nm)/Ni (160 nm)/Au (320 nm). The final samples were diced and encapsulated in TO39 packages using Ag nano-paste for cathode bonding and Au wire-contacting for the anode connection.

The X-ray diffraction (XRD) patterns were acquired in order to evaluate the Schottky contact quality. The measurements used a 9 kW Rigaku SmartLab diffractometer (Rigaku corp., Osaka, Japan)equipped with a monochromatic CuKα1 source that provided a wavelength (λ = 0.15406 nm). During the measurements, the incidence angle of the source (ω) was fixed to 0.5°, while the detector angle (2θ) scanned from 20° to 95°. Also, the incident slit was 0.1 mm, while the receiving slits were set to 20 mm.

X-ray photoelectron spectroscopy (XPS) measurements were performed using a polychromatic Al X-ray source at 13 kV with a power of 200 W. Vacuum was maintained at ~3 $\times$ 10^{-9} mbar. The energy analyzer was a 160 mm hemispherical type with a 1D detector (ASPECT, Sigma Surface Science) (Sigma Surface Science GmbH, Taunusstein, Germany). The diameter of the analysis spot was 1.3 mm. Prior to the XPS measurements, the samples were infrared-heated to ~100 °C for 5 min and etched by Ar sputtering at 0.5 keV for 10 min. The XPS spectra were analyzed by the CasaXPS software Version 2.3. 22PR1.0 using Shirley background determination and processed to remove the satellite peaks due to the K_β Al characteristic line. The binding energies were referenced to the C-C component in the C 1s signal at 284.8 eV associated with the adventitious carbon layer.

I-V characteristics of the packaged samples were measured at different temperatures between 60–500 K with a step of 20 K using a Keithley 4200 semiconductor (Keithley Corp, Cleveland, OH, USA) characterization system coupled with a Janis closed-cycle refrigerator system [14].

3. Results

3.1. Microstructural Investigations—X-ray Diffraction and X-ray Photoelectron Spectroscopy Analysis

The interfacial reaction of Pt and Cr with SiC was studied by XRD (Figure 1a,b) and XPS (Figure 1c,d).

Figure 1a evinces the presence of diffraction peaks at 2θ = 39.72°, 46.46°, 67.80°, 81.39°, and 85.59°. According to card no. 04-0802 of the ICDD (International Centre for Diffraction Data) database, these diffraction peaks can be attributed unambiguously to cubic Pt with a = 0.392 nm that belongs to the *Fm3m* space group. In addition, each Pt diffraction peak was accompanied at a lower 2θ = 39.17°, 45.56°, 66.34°, 79.66°, and 83.67° by a peak with a smaller intensity. These additional diffraction peaks could be assigned to strained Pt (Δ; Figure 1a) as a result of the relatively high temperature used for the sample preparation (~600 °C). According to the well-known Bragg law, lower 2θ values are associated to higher inter-planar distances. For instance, the inter-planar distances for different reflections increased as follows: from 0.226 nm to 0.230 nm (111), 0.195 nm to 0.199 nm (200), 0.138 nm to 0.141 nm (220), 0.118 nm to 0.120 nm (311), and 0.113 nm to 0.115 nm (222). Accordingly, the temperature induced the occurrence of a tensile lattice strain (ε) with different values along the atomic planes. Based on the calculated inter-planar distances, the lattice strain was: ~+1.7% (111), +2.1% (200), +2.2% (220), and +1.7% (311) and (222). The analysis also shows that no oxides or silicides were formed during the sample preparation.

In the case of *Cr/4H-SiC*, the XRD data show the presence of typical reflections of cubic Cr with a = 0.288 nm at 2θ = 44.39°, 64.58°, and 81.73°, respectively, as shown in Figure 1b. Unidentified diffraction peaks located at 2θ = 36.46°, 41.68°, 50.08°, 63.55°, and 79.40° could be assigned as (110), (113), (024), (214), and (306) reflections of Cr_2O_3 (ICDD, card no. 01-1294).

Further investigations related to the structure of the resulting compounds were performed by XPS for the *Pt 4f* and *Cr 2p* states. Figure 1c shows the high-resolution regions of *Pt 4f* in the range of the binding energies (BEs): 82 eV–68 eV. The BEs of these regions were calibrated using the binding energy of the adventitious carbon located at 284.8 eV.

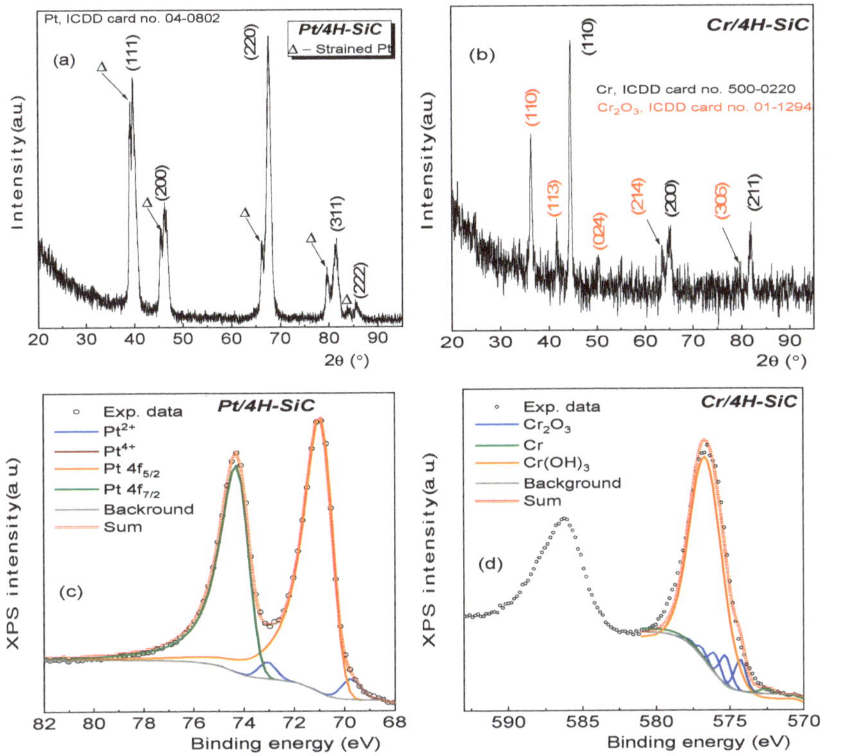

Figure 1. Grazing incidence XRD patterns for (**a**) *Pt/4H-SiC* and (**b**) *Cr/4H-SiC*. XPS spectra for (**c**) *Pt 4f* and (**d**) *Cr 2p* states with the corresponding fitting curves.

In the fitting process, a mixed Lorentzian–Gaussian function was used to identify the compounds from the investigated sample. The presence of Pt was confirmed by the analysis of the *Pt 4f* high resolution spectra exhibiting two components, one at 71.1 eV, attributed to $4f_{5/2}$ (orange), and one at 74.3 eV, associated with $4f_{7/2}$ (olive) peaks of Pt^0. The small asymmetry, observed in the *Pt 4f* peaks, was due to a small contribution of a peak at 69.3 eV in the $4f_{5/2}$ peak and a peak at 73.1 eV in the $4f_{7/2}$ peak associated with Pt^{2+} (blue line), which indicates that the Pt was superficially oxidized. Thus, the XPS data indicate that the Pt remained mostly in metallic form and that it did not form PtSi compounds. The results are in agreement with other studies conducted on Pt [26]. Larrieu et al. [27] investigated the evolution of *Pt 4f* with the annealing temperature, showing that the reaction of Pt to Pt_2Si or PtSi is characterized by peaks at BEs~72.5 eV, which were absent in our case. Furthermore, XPS analysis was performed to reveal the valence state of the Cr on the SiC. Based on the fitting of the experimental data, it is revealed that the Cr^{3+} species can be further divided into oxides, which showed a discrete multiplet structure, and hydroxides, which showed only a broad peak shape at BE = 576.7 eV. Also, Biesinger et al. [28] conducted in-depth XPS studies on Cr, showing a discrete multiplet structure, whereas the hydroxide gave only a broad peak shape.

Overall, the XRD analysis showed the formation of Pt as well as of a strained Pt layer. In addition, the XPS analysis indicated a superficial oxide at the surface. No other compounds were detected. On the other hand, in the case of Cr, the XRD and XPS analyses revealed a more complex structure resulting from the thermal treatment, which included Cr hydroxide and oxide compounds. Thus, a higher degree of inhomogeneity in the sample composition was expected.

3.2. *Electrical Characterization*

3.2.1. Temperature-Dependent Electrical Characteristics

Figure 2 shows the typical forward bias I-V characteristics of the fabricated samples in the temperature range of 60–500 K with a step of 40 K.

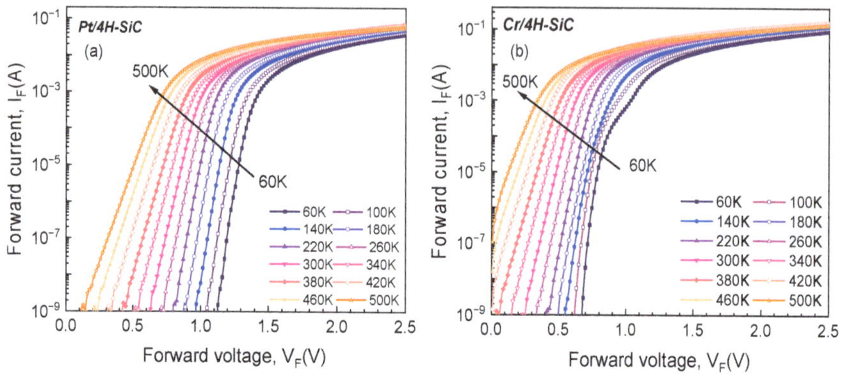

Figure 2. Experimental forward bias I-V characteristics of the SBDs at various temperatures: (**a**) *Pt/4H-SiC* sample; (**b**) *Cr/4H-SiC* sample.

Exponential behavior, covering at least six orders of magnitude, was identified for each experimental sample. For the *Pt/4H-SiC* sample, this dependence was evinced even at 500 K, while, for *Cr/4H-SiC*, the lower WF (and, consequently, lower SBH) led to a much higher increase in the saturation current with temperature.

3.2.2. Standard SBD Characterization

The I-V-T characteristics of an ideal SBD are governed by the thermionic emission (TE) law [29]:

$$I_F \cong I_S exp\left(\frac{V_F - I_F R_S}{nV_{th}}\right), \quad (1)$$

where R_S is the series resistance, n is the ideality factor, V_{th} = kT/q is the thermal voltage, and I_S is the saturation current,

$$I_S \cong A_n A_S T^2 exp\left(-\frac{\Phi_{Bn,T}}{V_{th}}\right), \quad (2)$$

where A_S is the designed contact area, A_n is the Richardson constant for electrons (146 A/K^2 cm^2 for n-type 4H-SiC), and $\Phi_{Bn,T}$ is the conventional Schottky barrier height.

The standard technique for extracting the main electrical parameters of an SBD entails the representation of ln (I_F) as a function of V_F, followed by linear fitting. The ideality factor and SBH are determined from the slope and intercept of this fit. For series resistance (R_S) determinations, a linear fit of the I-V plot in the high voltage domain is carried out. These electrical parameters, for the two fabricated samples, were extracted from the data in Figure 2. Their variation with temperature was plotted and is shown in Figure 3. According to the TE theory, an ideal Schottky contact yields a temperature-stable, constant SBH and ideality factor.

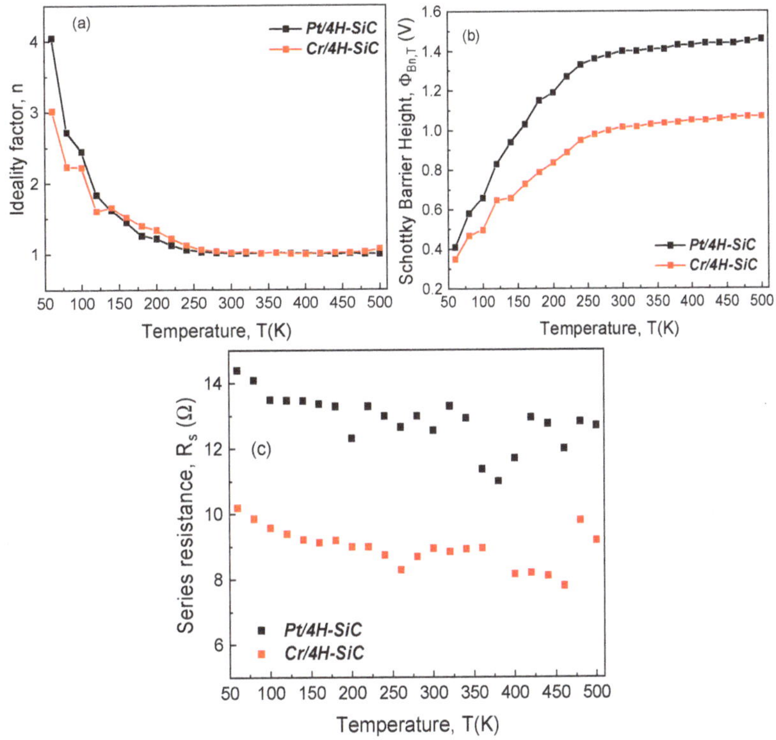

Figure 3. Temperature dependence of electrical parameters for the fabricated SiC SBD samples with Pt and Cr: (**a**) ideality factor; (**b**) Schottky barrier height; (**c**) series resistance.

Additionally, for SBDs with reasonably uniform contacts, the ideality factor should exhibit values close to unity. In our case, this situation corresponded to both the *Pt/4H-SiC* and *Cr/4H-SiC* samples only in the 260–500 K temperature interval. This range was also associated with a near-constant value for the Schottky barrier height. Conversely, a significant temperature dependence for these parameters was observed in the 60–240 K range. The behavior, coupled with the XRD findings, confirm that the devices' contacts were inhomogeneous.

3.2.3. State-of-the-Art Contact-Inhomogeneity Analysis

Multiple techniques were carried out in order to comprehensively evince the degree of contact inhomogeneity for the investigated diodes. Firstly, we evaluated the deviation from the ideal behavior, which is illustrated by the nkT vs. kT plot depicted in Figure 4. For this representation, the 340–500 K temperature range was considered, corresponding to an interval where the ideality factor was nearly constant.

Slight deviations from the ideal case ($n = 1$; green line in Figure 4) were observed. This anomaly is normally attributed to Schottky barrier non-uniformity [30]. In such cases, the ideality factor temperature dependence can be expressed as [31]:

$$n = 1 + \frac{T_0}{T}, \tag{3}$$

where $T_0 \neq 0$ is referred to as a T_0 anomaly [21]. A high value for T_0 corresponds to a higher degree of inhomogeneity. Values of 14.6 K for the *Pt/4H-SiC* sample and 50.1 K for the *Cr/4H-SiC* one were obtained from (3) after linear fitting was conducted on the curves illustrated in Figure 4. The relatively high value corresponding to the *Cr/4H-SiC* diode occurred mostly because of the data point at 500 K, indicating that this temperature level was beyond the normal capabilities for this type of contact. Excluding it from the analysis yielded T_0 = 20.8 K for *Cr/4H-SiC*, which was much more in agreement with its Pt counterpart. While performing this rudimentary T_0 anomaly technique can serve as a quick way to confirm that contact inhomogeneities do influence electrical characteristics significantly, it does not offer any physically relevant parameters.

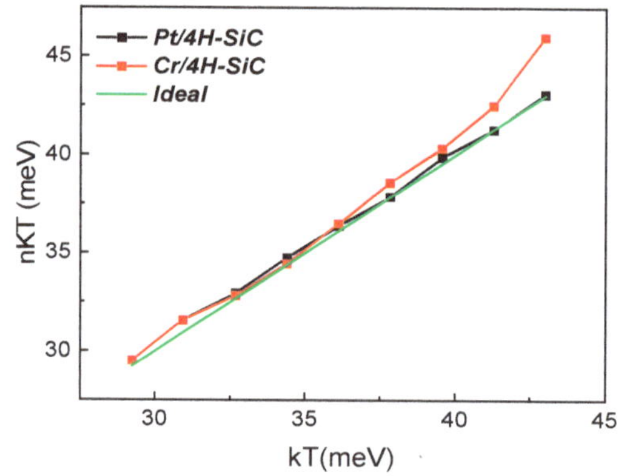

Figure 4. Plot of nkT vs. kT for both the fabricated samples. The ideal behavior (n = 1) is also reported as reference (green line).

A more thorough characterization can be performed using models that consider the well-known parallel conduction theory [21]. According to this, an experimental Schottky contact has numerous low-area regions ("patches") with independent barrier heights. If a Gaussian distribution of these patches is considered at the interface [32] with a mean Schottky carrier height (Φ^0_{Bn}) and standard deviation (σ), the conventional SBH temperature dependence (Figure 3b) can be expressed according to the following equation [32]:

$$\Phi_{Bn,T} = \Phi^0_{Bn} - \frac{q\sigma^2}{2kT}. \quad (4)$$

Representing $\Phi_{Bn,T}$ as a function of q/2kT (Figure 5), we can determine both Φ^0_{Bn} and σ from the resulting intercept and slope.

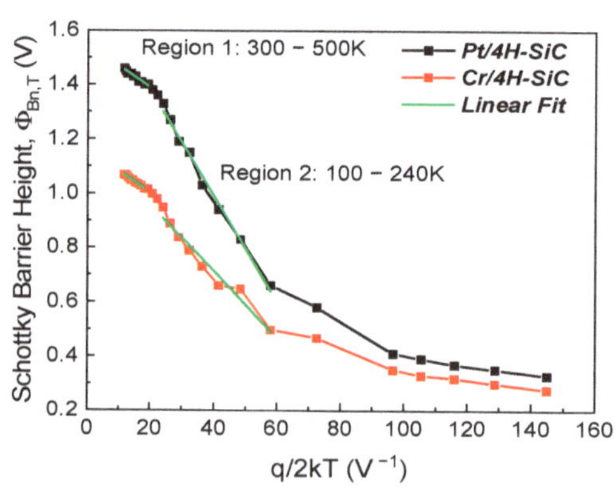

Figure 5. Conventional SBH vs. q/2kT.

Two linear regions can be identified on the graph in Figure 5, demonstrating that at least two Gaussian distributions were found on the contact's surface [33]. The extracted mean SBH and its standard deviation for the two temperature ranges are shown in Table 1 for both samples. The mean SBH for the *Pt/4H-SiC* sample was higher than that of the *Cr/4H-SiC* sample over both temperature intervals due to the difference in the metal WF values for the Schottky metals.

Table 1. Extracted values for the mean SBH and standard deviation for different temperature intervals.

Temperature Range, T (K)	Extracted Parameters			
	Pt/4H-SiC Sample		*Cr/4H-SiC* Sample	
	Φ_{Bn}^{0} (V)	σ (V)	Φ_{Bn}^{0} (V)	σ (V)
100–240	1.774	0.134	1.206	0.111
300–500	1.544	0.088	1.163	0.089

Note that a third region can theoretically be identified for the 60–100 K range. At such low temperature levels, however, conventionally extracted SBH values are significantly affected by errors stemming from the bias interval window of analysis and, possibly, carrier freeze-out [34,35]. Since the Gaussian distribution model does not consider such effects, no significance can be derived from applying the technique to the 60–100 K measurements.

The discrepancies between the values determined using this technique and their counterparts from the conventional method (see Figure 3b) make this analysis unable to explain the overall behavior of our fabricated samples. As we can see in Table 1, the mean SBH presented higher values over both temperature intervals than what the conventional SBH trend would suggest (see Figure 3b). Practically, this means that the determined Φ_{Bn}^{0} was not the dominant one.

Since the behavior of the SBD samples was closer to the ideal ($n < 1.07$) over 300–500 K, this higher temperature interval will be referenced further in our analysis of the contact inhomogeneity.

Additional information about the impact of the contact inhomogeneity is given by the Richardson plot [31], from which both an effective SBH $\left(\Phi_{Bn-eff}\right)$ and active area ($A_{S\text{-eff}}$) can be determined. The Richardson plots for our samples are governed by the expression:

$$ln\left(\frac{I_F}{T^2}\right) = ln(A_S A^*) - \frac{q\left(\Phi_{Bn-eff} - \frac{V_F}{n_{med}}\right)}{kT}, \quad (5)$$

where n_{med} represents the mean ideality factor value over a certain temperature interval.

Using the standard deviation values determined before (see Table 1), we can construct a modified Richardson plot,

$$ln\left(\frac{I_F}{T^2}\right) - \left(\frac{q^2\sigma^2}{2k^2T^2}\right) = ln(A_S A^*) - \frac{q\left(\Phi_{Bn-eff} - \frac{V_F}{n_{med}}\right)}{kT}, \quad (6)$$

where σ is the standard deviation determined using the conventional SBH vs. q/2kT plot from Figure 5 with the values in Table 1 for both temperature intervals: low (100–240 K) and high (300–500 K). Since n_{med} is truly representative for measurements only in the high-temperature region, we constructed Richardson plots by taking several bias voltages from the characteristics only in this domain, which was further restricted to 340–500 K. This approach was taken in order to ensure that all the selected curves exhibited exponential behavior for each V_F. The voltage interval of 0.5–0.75 V was identified for *Pt/4H-SiC* and 0.2–0.4 V for *Cr/4H-SiC*. The effective SBH was determined from these intervals, and the optimum voltage for constructing both the Richardson and modified Richardson plots was selected such that fitting errors were minimized. Thus, a bias voltage of 0.6 V was chosen for the *Pt/4H-SiC* sample and 0.25 V for *Cr/4H-SiC*. The plots are given in Figure 6.

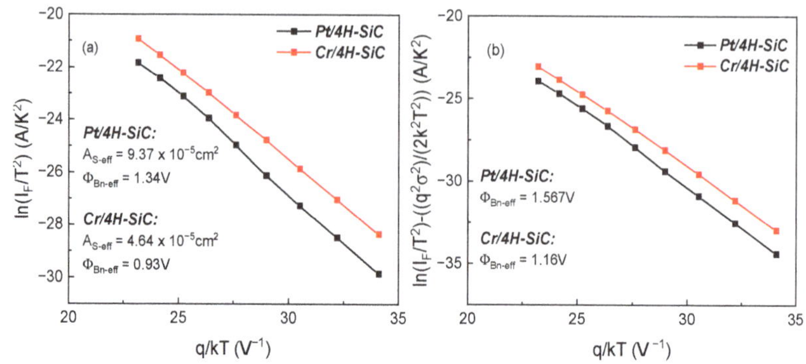

Figure 6. (**a**) Optimal Richardson plot; (**b**) optimal modified Richardson plot, using standard deviation from Table 1 (300–500 K).

The effective active area (A_{S-eff}) of the samples was determined as being one order of magnitude smaller than the designed one (~12.56 × 10^{-4} cm^2). This is another probative indicator of the occurrence of inhomogeneity on the contact surface. Moreover, the *Cr/4H-SiC* sample exhibited an A_{S-eff} two times lower than *Pt/4H-SiC*, indicating a higher degree of inhomogeneity. Accordingly, the current flow through the device is favored by the small, low-barrier patches located on the contact surface. The effective SBH value, determined using the standard Richardson plot, is also given in Figure 6. It better corresponded to the electrical behavior of the fabricated samples, as it was in suitable agreement with the variation in the trend of the conventional SBH values plotted in Figure 3b. Conversely, the SBH value obtained from the modified Richardson plot naturally mimicked the one determined using the Werner and Güttler plot (see Figure 5). It is, theoretically, the extrapolation at infinite temperature of the governing barrier height. Hence, in practice, this Gaussian distribution method does not produce relevant parameters for highly inhomogeneous devices. The temperature threshold after which these extracted barrier heights would become indicative of actual device current flow far exceeds operational levels.

Finally, for a complete elucidation of the fabricated samples' electrical behavior over the entire investigated domain, a more comprehensive approach was undertaken. It was based on our developed *p-diode* model [24], which also uses the parallel conduction theory as the underlying principle. According to it, a real Schottky contact behaves like a grouping of multiple parallel-connected near-ideal diodes (n capped at 1.03), each with its specific barrier height and non-uniformity parameter (p_{eff}, giving a quantitative depiction of the occupied area proportion). Distinctively from the Gaussian approach, each parallel diode is also associated with a series resistance that can limit its current contribution to the overall I_F as the bias increases. The forward curves of both the *Pt/4H-SiC* and *Cr/4H-SiC* samples were characterized with the *p-diode* technique over the entire 60–500 K range. The model-fitted curves are depicted in Figure 7.

An excellent agreement between the fitted curves and experimental measurements can be observed, even at low temperatures, where contact inhomogeneity determines strong deviations from the exponential I_F–V_F dependence. Modeling at such near-cryogenic levels is possible distinctly because of the *p-diode*'s consideration of patch resistive effects (ignored in the Gaussian distribution model). Four parallel diodes (Dp1–Dp4, Figure 8) were necessary to fully replicate the electrical behavior for both samples in the encompassing temperature span, corresponding to different barrier zones. The obtained parameters are given in Table 2. These regions were associated with the different compounds on the contact surface evinced by the XRD and XPS analyses.

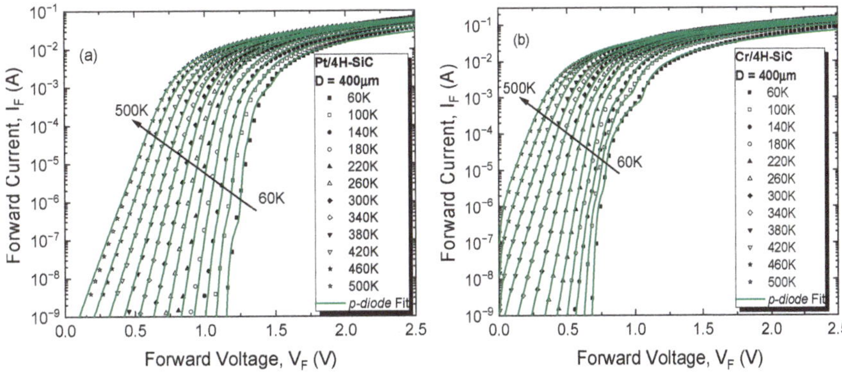

Figure 7. *p-diode* model-fitted curves for 4H-SiC Schottky diode samples with (**a**) Pt and (**b**) Cr metals.

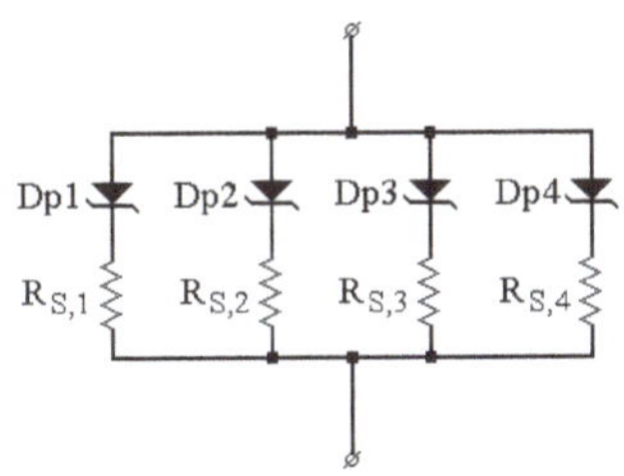

Figure 8. *p-diode* model equivalent schematic for 4H-SiC Schottky diode samples with Pt and Cr metals.

Table 2. Extracted *p-diode* model parameters.

Sample	Parallel Diodes	Φ_{Bn} [V]	n	p_{eff}	R_S [Ω]
Pt/4H-SiC	Pt-Dp1	1.51	1.03	0.51	44–60
	Pt-Dp2	1.35		3.04	150–48
	Pt-Dp3	1.3		4.39	150–800
	Pt-Dp4	1.21		9.26	200 k–40 k
Cr/4H-SiC	Cr-Dp1	1.1		0.37	14.2–21.9
	Cr-Dp2	0.93		3.73	50–430
	Cr-Dp3	0.8		10.98	300–550
	Cr-Dp4	0.73		13.39	~15 k

For the *Pt/4H-SiC* sample, Pt-Dp1, covering most of the overall contact surface, exhibited the highest barrier height and, at low temperatures, was only influential at a high bias. As the temperature increased, this parallel diode began to contribute significant current over the entire V_F range. Conversely, Pt-Dp4 was only prominent in the low-temperature, low-bias regions, with its impact becoming negligible at higher T and V_F levels. Parallel diodes Pt-Dp2 and Pt-Dp3 had comparable influence on the forward characteristics throughout the entire bias and temperature intervals. Their lumped contributions were especially relevant in the 300–500 K domain, as also confirmed by the Φ_{Bn} values similar to that obtained from the Richardson plot (see Figure 6).

For the *Cr/4H-SiC* sample, Cr-Dp2 was the main current contributor, which was once again verified by the results obtained from the Richardson plots. Cr-Dp3 and Cr-Dp4 were responsible for the current flow in the lower ranges of bias and temperature, while Cr-Dp1 mostly influenced I_F at the top end of the temperatures at high V_F.

The *p-diode* analysis completely explains the forward electrical behavior of the investigated samples throughout the entire temperature domain. Both exhibited a considerable degree of contact inhomogeneity, with multiple current paths becoming preferential as the bias and thermal conditions evolved. Even so, suitable I_F levels can be found, where the samples essentially behaved like ideal Schottky diodes, enabling their use in temperature and gas sensing applications. For such uses, the Pt samples are more desirable due to their overall larger barrier height in order to obtain both a higher sensitivity and a wider operable temperature range [36].

An important conclusion of the comprehensive inhomogeneity analysis is that all of the employed techniques, apart from the *p-diode* method, either analyzed the Schottky contact area as a whole or required restricted temperature intervals in order to produce trustworthy results. While being useful tools for preliminary contact quality diagnosis, they must be accompanied by *p-diode* modeling in order to accurately and completely assess the forward electrical behavior of such wide-temperature-range SiC diodes.

4. Conclusions

This paper analyzed the contact inhomogeneity of Pt/SiC and Cr/SiC. The fabricated samples were subjected to XRD and XPS analyses, revealing possible inhomogeneity sources. For the *Pt/4H-SiC* sample, slight traces of oxides and a strained layer were identified. In the case of *Cr/4H-SiC*, more pronounced inhomogeneity was evinced, stemming from hydroxide and oxide compounds. The conventional electrical characterization demonstrated important variations in the barrier height and ideality factor with temperature, which confirmed the formation of a non-uniform contact. Subsequent inhomogeneity modeling employed techniques of gradually increasing complexity, which confirmed that the *Cr/4H-SiC* diode was more affected by this spurious influence.

The forward characteristics of both samples were completely modeled with our *p-diode* technique over the entire investigated domain. Each of the samples behaved essentially as four parallel-connected ideal diodes, with variable influence on the current conduction, depending on the temperature and bias levels. The *Pt/4H-SiC* diode's current was mainly given by a contact region with a barrier of 1.3–1.35 V. For the *Cr/4H-SiC* sample, a main barrier of 0.93 V governed the current conduction. Both of these results were corroborated by the ones obtained from the Richardson plots and conventional SBH extraction in the plateaus corresponding to the high-temperature domain.

Employing the *p-diode* modeling was crucial in order to identify the suitable operable bias and temperature conditions for these samples.

Author Contributions: Conceptualization, R.P. and G.B.; methodology, R.P., G.P. and C.R.; software, D.-T.O.; validation, D.-T.O., F.D. and E.C.; formal analysis, D.-T.O.; investigation, R.P., C.R., N.D., A.E., F.D. and A.M.I.; resources, R.P., G.P., G.B., N.D. and A.M.I.; data curation, R.P. and G.P.; writing—original draft preparation, R.P., G.P., C.R. and A.E.; writing—review and editing, G.B. and F.D.; visualization, F.D., A.E. and E.C.; supervision, R.P., G.P. and G.B.; project administration, R.P., G.P., G.B., N.D. and A.M.I.; funding acquisition, R.P., G.P., G.B., N.D. and A.M.I. All authors have read and agreed to the published version of the manuscript.

Funding: This work was supported by a grant of the Ministry of Research, Innovation, and Digitization, CCCDI—UEFISCDI, project numbers PN-III-P2-2.1-PED-2021-2688, PN-III-P1-1.1-TE-2021-0231, and PN-III-P2-2.2-PED-2021-4158, within PNCDI III, and by a grant from the National Program for Research of the National Association of Technical Universities, GNAC ARUT 2023, project number 137. Also, N.D. acknowledges the support by the contract PN 23.21.01.06 sponsored by the Romanian Ministry of Research, Innovation, and Digitalization.

Data Availability Statement: Data are contained within the article.

Conflicts of Interest: The authors declare no conflicts of interest.

References

1. Godignon, P.; Jorda, X.; Vellvehi, M.; Perpina, X.; Banu, V.; Lopez, D.; Barbero, J.; Brosselard, P.; Massetti, S. SiC Schottky diodes for harsh environment space applications. *IEEE Trans. Ind. Electron.* **2011**, *58*, 2582–2589. [CrossRef]
2. She, X.; Huang, A.Q.; Lucia, O.; Ozpineci, B. Review of Silicon Carbide Power Devices and Their Applications. *IEEE Trans. Ind. Electron.* **2017**, *64*, 8193–8205. [CrossRef]

3. Draghici, F.; Brezeanu, G.; Pristavu, G.; Pascu, R.; Badila, M.; Pribeanu, A.; Ceuca, E. 400 °C Sensor Based on Ni/4H-SiC Schottky Diode for Reliable Temperature Monitoring in Industrial Environments. *Sensors* **2019**, *19*, 2384. [CrossRef] [PubMed]
4. Somma, R.; Troise, C.; Zeni, L.; Minardo, A.; Fedele, A.; Mirabile, M.; Natale, G. De Long-Term Monitoring with Fiber Optics Distributed Temperature Sensing at Campi Flegrei: The Campi Flegrei Deep Drilling Project. *Sensors* **2019**, *19*, 1009. [CrossRef] [PubMed]
5. Hermans, T.; Nguyen, F.; Robert, T.; Revil, A. Geophysical Methods for Monitoring Temperature Changes in Shallow Low Enthalpy Geothermal Systems. *Energies* **2014**, *7*, 5083–5118. [CrossRef]
6. Qi, J.; Yang, X.; Li, X.; Chen, W.; Tian, K.; Wang, M.; Guo, S.; Yang, M. Comparative Temperature Dependent Evaluation and Analysis of 1.2-kV SiC Power Diodes for Extreme Temperature Applications. *IEEE Trans. Power Electron.* **2020**, *35*, 13384–13399. [CrossRef]
7. Kim, S.; Choi, J.; Jung, M.; Joo, S.; Kim, S. Silicon carbide-based hydrogen gas sensors for high-temperature applications. *Sensors* **2013**, *13*, 13575–13583. [CrossRef]
8. Pascu, R.; Craciunoiu, F.; Pristavu, G.; Brezeanu, G.; Kusko, M. Oxide trap states versus gas sensing in SiC-MOS capacitors—The effect of N- and P- based post oxidation processes. *Sens. Actuators B Chem.* **2017**, *245*, 911–922. [CrossRef]
9. Taylor, N.R.; Yu, Y.; Ji, M.; Aytug, T.; Mahurin, S.; Mayes, R.; Cetiner, S.; Paranthaman, M.P.; Ezell, D.; Cao, L.R.; et al. Thermal and radiation response of 4H-SiC Schottky diodes with direct-write electrical contacts. *Appl. Phys. Lett.* **2020**, *116*, 252108. [CrossRef]
10. Della Corte, F.G.; Pangallo, G.; Rao, S.; Carotenuto, R.; Iero, D.; Merenda, M.; Pezzimenti, F. Use of 4H-SiC-based diodes as temperature sensors. In Proceedings of the 2019 International Semiconductor Conference (CAS), Sinaia, Romania, 9–11 October 2019; pp. 71–74. [CrossRef]
11. Zhang, N.; Lin, C.; Senesky, D.G.; Pisano, A.P. Temperature sensor based on 4H-silicon carbide pn diode operational from 20 °C to 600 °C. *Appl. Phys. Lett.* **2014**, *104*, 073504. [CrossRef]
12. Rao, S.; Pangallo, G.; Pezzimenti, F.; Corte, F.G.D. High-Performance Temperature Sensor Based on 4H-SiC Schottky Diodes. *IEEE Electron Device Lett.* **2015**, *36*, 720–722. [CrossRef]
13. Lebedev, A.A.; Kozlovski, V.V.; Davydovskaya, K.S.; Levinshtein, M.E. Radiation Hardness of Silicon Carbide upon High-Temperature Electron and Proton Irradiation. *Materials* **2021**, *14*, 4976. [CrossRef] [PubMed]
14. Pascu, R.; Pristavu, G.; Oneata, D.T.; Stoian, M.; Romanitan, C.; Kusko, M.; Draghici, F.; Brezeanu, G. Enhanced Method of Schottky Barrier Diodes Performance Assessment. *Rom. J. Inf. Sci. Technol.* **2023**, *2023*, 181–192. [CrossRef]
15. Li, D.; Li, X.; Chang, G.; Qi, F.; Packwood, M.; Pottage, D.; Wang, Y.; Luo, H.; Dai, X.; Liu, G. Characterization of A 3.3 kV Si—SiC Hybrid Power Module in Half—Bridge Topology for Traction Inverter Application. *IEEE Trans. Power Electron.* **2020**, *35*, 13429–13440. [CrossRef]
16. Kyoung, S.; Jung, E.S.; Sung, M.Y. Post-annealing processes to improve inhomogeneity of Schottky barrier height in Ti/Al 4H-SiC Schottky barrier diode. *Microelectron. Eng.* **2016**, *154*, 69–73. [CrossRef]
17. Bourenane, K.; Keffous, A.; Nezzal, G. Electrical properties of Schottky diode Pt/SiC and Pt/porous SiC performed on highly resistif p-type 6H-SiC. *Vacuum* **2007**, *81*, 663–668. [CrossRef]
18. Saxena, V.; Nong, J.; Steckl, A.J. High-Voltage Ni- And Pt-SiC Schottky Diodes Utilizing Metal Field Plate Termination. *IEEE Trans. Electron Devices* **1999**, *46*, 456–464. [CrossRef]
19. Khanna, S.; Noor, A.R. Electrical Characterization of Chromium/4H-SIC Schottky Barrier Diodes. *Int. J. Eng. Sci. Technol.* **2010**, *2*, 220–226.
20. Strel'Chuk, A.M.; Kalinina, E.V. Schottky diodes based on 4H-SiC epitaxial layers. *J. Phys. Conf. Ser.* **2021**, *2103*, 012235. [CrossRef]
21. Tung, R.T. Recent advances in Schottky barrier concepts. *Mater. Sci. Eng. R Rep.* **2001**, *35*, 1–138. [CrossRef]
22. Kumar, V.; Verma, J.; Maan, A.S.; Akhtar, J. Epitaxial 4H–SiC based Schottky diode temperature sensors in ultra-low current range. *Vacuum* **2020**, *182*, 109590. [CrossRef]
23. Cowley, A.M. Titanium-silicon Schottky barrier diodes. *Solid State Electron.* **1970**, *13*, 403–414. [CrossRef]
24. Brezeanu, G.; Pristavu, G.; Draghici, F.; Pascu, R.; Della Corte, F.; Rascuna, S. Enhanced Non-Uniformity Modeling of 4H-SiC Schottky Diode Characteristics Over Wide High Temperature and Forward Bias Ranges. *IEEE J. Electron Devices Soc.* **2020**, *8*, 1339–1344. [CrossRef]
25. Pascu, R.; Pristavu, G.; Brezeanu, G.; Draghici, F.; Godignon, P.; Romanitan, C.; Serbanescu, M.; Tulbure, A. 60–700 K CTAT and PTAT Temperature Sensors with 4H-SiC Schottky Diodes. *Sensors* **2021**, *21*, 942. [CrossRef] [PubMed]
26. Mazzotta, E.; Rella, S.; Turco, A.; Malitesta, C. XPS in development of chemical sensors. *RSC Adv.* **2015**, *5*, 83164–83186. [CrossRef]
27. Larrieu, G.; Dubois, E.; Wallart, X.; Baie, X. Formation of Pt-based silicide contacts: Kinetics, stochiometry and current drive capabilities. *J. Appl. Phys.* **2003**, *94*, 7801–7810. [CrossRef]
28. Biesinger, M.C.; Payne, B.P. Resolving surface chemical states in XPS analysis of first row transition metals, oxides and hydroxides: Cr, Mn, Fe, Co and Ni. *Appl. Surf. Sci.* **2011**, *257*, 2717–2730. [CrossRef]
29. Gammon, P.M.; Pérez-Tomás, A.; Shah, V.A.; Vavasour, O.; Donchev, E.; Pang, J.S.; Myronov, M.; Fisher, C.A.; Jennings, M.R.; Leadley, D.R.; et al. Modelling the inhomogeneous SiC Schottky interface. *J. Appl. Phys.* **2013**, *114*, 223704. [CrossRef]
30. Sullivan, J.P.; Tung, R.T.; Pinto, M.R.; Graham, W.R. Electron transport of inhomogeneous Schottky barriers: A numerical study. *J. Appl. Phys.* **1991**, *70*, 7403–7424. [CrossRef]
31. Roccaforte, F.; Giannazzo, F.; Alberti, A.; Spera, M.; Cannas, M.; Cora, I.; Pécz, B.; Iucolano, F.; Greco, G. Barrier inhomogeneity in vertical Schottky diodes on free standing gallium nitride. *Mater. Sci. Semicond. Process.* **2019**, *94*, 164–170. [CrossRef]

32. Werner, J.H.; Güttler, H.H. Barrier inhomogeneities at Schottky contacts. *J. Appl. Phys.* **1991**, *69*, 1522–1533. [CrossRef]
33. Gülnahar, M. Temperature dependence of current-and capacitance-voltage characteristics of an Au/4H-SiC Schottky diode. *Superlattices Microstruct.* **2014**, *76*, 394–412. [CrossRef]
34. Gammon, P.M.; Fisher, C.A.; Shah, V.A.; Jennings, M.R.; Pérez-Tomás, A.; Burrows, S.E.; Myronov, M.; Leadley, D.R.; Mawby, P.A. The cryogenic testing and characterisation of SiC diodes. *Mater. Sci. Forum* **2014**, *778–780*, 863–866. [CrossRef]
35. Triendl, F.; Pfusterschmied, G.; Pobegen, G.; Konrath, J.P.; Schmid, U. Theoretical and experimental investigations of barrier height inhomogeneities in poly-Si/4H-SiC heterojunction diodes. *Semicond. Sci. Technol.* **2020**, *35*, 115011. [CrossRef]
36. Pristavu, G.; Brezeanu, G.; Badila, M.; Draghici, F.; Pascu, R.; Craciunoiu, F.; Rusu, I.; Pribeanu, A. Barrier Stability of Pt/4H-SiC Schottky Diodes Used for High Temperature Sensing. *Mater. Sci. Forum* **2017**, *897*, 606–609. [CrossRef]

materials

MDPI

Article

Characteristics of the Discoloration Switching Phenomenon of 4H-SiC Single Crystals Grown by PVT Method Using ToF-SIMS and Micro-Raman Analysis

Seul-Ki Kim [1,2], Hajun Kim [3], Hyun Sik Kim [4], Tae Eun Hong [5], Younki Lee [3,*] and Eun Young Jung [6,*]

[1] Semiconductor Materials Center, Korea Institute of Ceramic Engineering and Technology, Jinju 52851, Republic of Korea; skkim@kicet.re.kr
[2] Department of Materials Engineering, Gyeongsang National University, Jinju 52828, Republic of Korea
[3] Department of Materials Engineering and Convergence Technology, Gyeongsang National University, Jinju 52828, Republic of Korea; khj42021@gmail.com
[4] Analysis and Standards Center, Korea Institute of Ceramic Engineering & Technology, Jinju 52851, Republic of Korea; hyunkim@kicet.re.kr
[5] Division of High-Technology Materials Research, Korea Basic Science Institute, Busan 46742, Republic of Korea; tehong@kbsi.re.kr
[6] The Institute of Electronic Technology, College of IT Engineering, Kyungpook National University, Daegu 41566, Republic of Korea
* Correspondence: ylee@gnu.ac.kr (Y.L.); eyjung@knu.ac.kr (E.Y.J.)

Abstract: The discoloration switching appearing in the initial and final growth stages of 4H-silicon carbide (4H-SiC) single crystals grown using the physical vapor transport (PVT) technique was investigated. This phenomenon was studied, investigating the correlation with linear-type micro-pipe defects on the surface of 4H-SiC single crystals. Based on the experimental results obtained using time-of-flight secondary ion mass spectrometry (ToF-SIMS) and micro-Raman analysis, it was deduced that the orientation of the 4H-SiC c-axis causes an axial change that correlates with low levels of carbon. In addition, it was confirmed that the incorporation of additional elements and the concentrations of these doped impurity elements were the main causes of discoloration and changes in growth orientation. Overall, this work provides guidelines for evaluating the discoloration switching in 4H-SiC single crystals and contributes to a greater understanding of this phenomenon.

Keywords: 4H-SiC single crystal; physical vapor transport (PVT); discoloration switching; ToF-SIMS; Raman spectroscopy

Citation: Kim, S.-K.; Kim, H.; Kim, H.S.; Hong, T.E.; Lee, Y.; Jung, E.Y. Characteristics of the Discoloration Switching Phenomenon of 4H-SiC Single Crystals Grown by PVT Method Using ToF-SIMS and Micro-Raman Analysis. *Materials* **2024**, *17*, 1005. https://doi.org/10.3390/ma17051005

Academic Editors: Marilena Vivona and Mike Jennings

Received: 16 January 2024
Revised: 12 February 2024
Accepted: 20 February 2024
Published: 22 February 2024

1. Introduction

Silicon carbide (SiC) has attracted significant attention in recent years owing to its outstanding electrical, mechanical, and thermal properties. As a result, SiC is commonly employed in many fields due to its incorporation into power devices and optoelectronics [1–5]. Over 200 SiC polytypes are known to exist, with cubic (3C-SiC) and hexagonal (4H-SiC or 6H-SiC) —modified compounds being the most commonly used; these structural differences significantly influence the electrical properties and potential applications of each polytype. Among the various SiC polytypes, the cubic (3C-SiC) and hexagonal (4H-SiC or 6H-SiC) structures are the most commonly studied and used [2,6,7].

For example, 4H-SiC is known to exhibit a high-breakdown electric field at room temperature, in addition to high thermal conductivity and a wide bandgap [8,9]. A number of SiC polytypes also exist that exhibit high process temperatures and chemical stabilities, among which 4H-SiC is the most widely used in power electronic applications [10]. Commercial SiC may be classified as either conductive SiC or high-purity semi-insulating SiC. In the case of high-purity semi-insulating SiC, almost no impurity dopant atoms are present, thereby accounting for the colorless nature of 4H-SiC, which is treated to remove all impurities. More specifically, commercial 4H-SiC is colorless due to its wide band gap, which prevents absorption over the wavelength range of visible light. However, upon doping conductive SiC with N, P, Al, or B to generate additional energy levels [11], absorption occurs in the visible light region. In addition, doped SiC crystals are known to exhibit unique color characteristics, wherein the SiC crystal color becomes darker upon increasing the concentration of impurity elements [12]. Moreover, 4H-SiC single crystals are mainly grown using the slow physical vapor transport (PVT) approach [13,14], which leads to different crystal colors in the initial and final states of crystal growth. With these considerations in mind, it is clearly important to establish an analytical approach for monitoring this color change and to determine the mechanism behind this phenomenon in 4H-SiC crystals under different impurity concentrations. This is of particular importance in the context of process control and failure analysis for the obtained SiC crystals.

Thus, to investigate the structural properties and spatial distributions involved in the discoloration of 4H-SiC single crystals, discoloration switching is performed, and the formation of linear-type micro-pipe defects is evaluated during two different growth periods, namely the initial (Case I) and final (Case II) growth periods. Considering these two growth periods, the characteristics of the 4H-SiC single crystals obtained by PVT are examined using gas discharge mass spectrometry (GDMS), micro-Raman spectroscopy, X-ray diffraction (XRD), photoluminescence (PL) spectroscopy, magnetic sector secondary-ion mass spectrometry (magnetic sector-SIMS), and time-of-flight secondary ion mass spectrometry (ToF-SIMS).

2. Materials and Methods

2.1. Preparation of 3C-SiC as a Precursor Material

The 3C-SiC (cubic silicon carbide) powder was synthesized in a customized furnace via a chemical vapor deposition (CVD) process, which included vaporization, pyrolysis, nucleation, oxidation–reduction, and substitution. Subsequently, it was reacted with granules at a high temperature [15]. The precursor gas comprised commercial methyltrichlorosilane (MTS) as the silicon precursor, as well as ammonia (NH_3) and carbon dioxide (CO_2). Ethylene (C_2H_4) and propane (C_3H_8) were used as hydrocarbon precursors due to their easy decomposition ability under high-temperature conditions.

A mixture of H_2 and Ar was employed as the carrier gas. The color of the synthesized 3C-SiC was black, indicating the presence of impurities or defects in the material. The 3C-SiC powder was crushed and milled using specially developed equipment to minimize contamination. The residual acid salt was removed by volatilization, and any excess oxygen present in the SiC was removed through denitrification under heating at 105 °C under reduced pressure in an argon environment. All other impurities present in the synthesized SiC were eliminated through the decarburization method by heat treatment at 850 °C for 1 h under oxygen atmosphere conditions to give a dried 3C-SiC powder with a purity of >90% [15,16]. The method for removing excess oxygen present in 3C-SiC powder was explained in detail as follows. There are some impurities in the 3C-SiC powder synthesized by the CVD process, such as excess oxygen, free carbon, and free silicon compounds (F-Si or F-SiO_2). These by-products were removed through acid leaching, decarburization, and esterification. Firstly, the free silicon compounds were eliminated through acid leaching purification by using the mixed hydrochloric acid solution with a sodium catalyst. Thereafter, in the case of decarburization purification, the 3C-SiC powder was treated through thermal treatment by passing oxygen gas in order to remove CO_2.

Then, the residual oxygen was purified by the OH substitution of acid through esterification at 105 °C on a hot plate [15,16]. Subsequently, the obtained powder was sieved through a 100 µm mesh to remove the fine and coarse particles. The synthesized 3C-SiC powder was also subjected to chemical pretreatment to remove any impurities introduced during the synthesis.

The elemental composition of the 3C-SiC powder was obtained using gas discharge mass spectrometry (GDMS), as listed in Table 1, and a purity of 99.999% was confirmed.

The XRD patterns of the 3C-SiC powder showed peaks corresponding to 3C-SiC (β-SiC phase), as shown in Figure 1. The peaks at 35.7°, 41.2°, 59.9°, and 71.6° are attributed to the (111), (200), (220), and (311) planes of the β-SiC phase, respectively [15–18].

Table 1. Detailed element compositions of the 3C-SiC powder and the grown 4H-SiC single crystals, as determined using gas discharge mass spectrometry (GDMS).

Element (ppm)	B	Al	P	Na	V	Ti	Fe	Ca	Cl	S
3C-SiC raw powder	1.10	0.08	<0.05	0.16	<0.05	<0.05	0.34	<0.50	1.1	0.14
4H-SiC crystal	1.20	0.10	<0.05	0.10	<0.05	<0.05	<0.10	<0.50	0.71	0.07

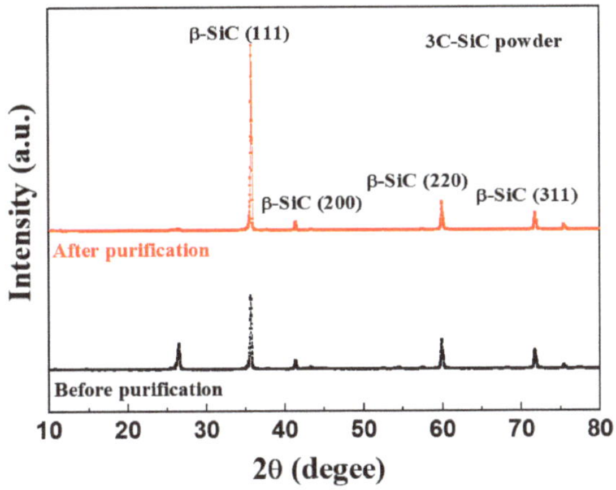

Figure 1. X-ray diffraction (XRD) patterns of the purified 3C-SiC powder as a starting material for 4H-SiC single crystals.

2.2. Growth of 4H-SiC Single Crystals Using the PVT Process

The 4H-SiC single crystals were grown from the 3C-SiC source powder using PVT [15], as outlined in Figures 2a and 3. To grow 4H-SiC single crystals, the 3C-SiC powder was placed on the graphite crucible in the PVT reactor chamber. Crystal growth was conducted under an inert gas environment, wherein a 4H-SiC crystal ingot was grown by sublimation of the powder in the heated PVT reactor crucible. Detailed experimental conditions for the growth of the 4H-SiC single crystals using PVT as shown in Table 2.

Table 2. Detailed experimental conditions for the growth of the 4H-SiC single crystals using PVT.

Precursor material	3C-SiC polycrystalline powder
Reaction gas	Mixture of argon and nitrogen gas
Working pressure	35 Torr
Growth temperature	2300 °C
Growth rate	1400 μm/h
Growth time	30 h
Two case studies	Case I: Bright sample Case II: Dark sample

More specifically, the growth of a single crystal was conducted under high-temperature conditions of 2300 °C and a pressure of 35 Torr in the presence of an argon/nitrogen gas mixture (99.96% purity). The injected argon gas flow rate was controlled using a mass flow meter. To maintain a stable gas pressure of 35 Torr during crystal growth, a mechanical pump was adjusted according to the required argon gas flow rate. The seed lid was attached to the top of the graphite crucible containing the 4H-SiC substrate. Crystal growth was allowed to proceed over 30 h with an average growth rate of ~1400 μm/h. After completion of the growth and cooling stages, the 4H-SiC crystal ingot was detached from the graphite crucible and sliced parallel to the c-axis direction. Subsequently, the sliced 4H-SiC was polished and cut perpendicular to the c-axis for analysis. Figure 2b shows a cross-sectional diagram of the 4H-SiC single crystal's growth. The elemental composition of the grown 4H-SiC single crystals was obtained using GDMS (see Table 1), and a purity of 99.999% was confirmed.

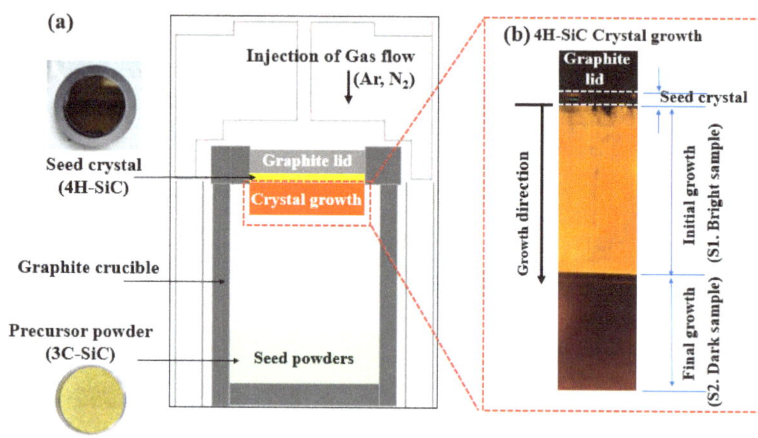

Figure 2. (**a**) Experimental set-up for the PVT process, and (**b**) a cross-sectional diagram for 4H-SiC single crystal growth.

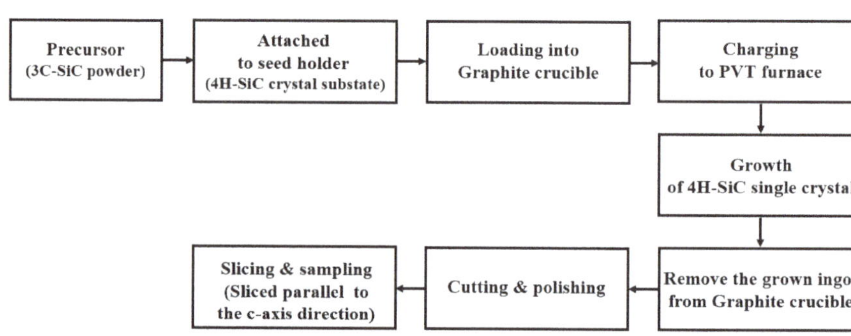

Figure 3. Detailed experimental procedure for growth of the 4H-SiC single crystals using the PVT process.

Figure 4a,b presents photographic and optical images (under light-emitting diode (LED) illumination), respectively, recorded for the surface of a sliced 4H-SiC single crystal. Under LED illumination, the sliced 4H-SiC single crystal appeared to be divided into two color regions, namely Case I (bright, near side) and Case II (dark, external side). Detailed experimental conditions and parameters for the identification of the sliced 4H-SiC single crystals are presented in Table 3.

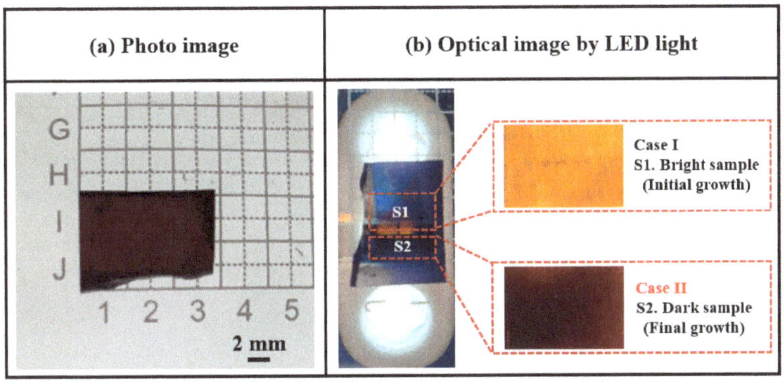

Figure 4. (**a**) Photographic image of the sliced 4H-SiC single crystal, and (**b**) optical images recorded under LED illumination for cases I and II.

Table 3. Detailed experimental conditions and identification parameters for the sliced 4H-SiC single crystal sample regions corresponding to cases I and II.

Sample Identification	Condition	Collection Position	Photographic Image
Case I	S1. Bright sample	Seed near side of 4H-SiC Ingot	
Case II	S2. Dark sample	External side of 4H-SiC Ingot	

2.3. Analysis and Characterization

The elemental compositions of the prepared 3C-SiC powder and the grown 4H-SiC single crystal were determined using gas discharge mass spectrometry (GDMS; Element GD Plus GD-MS, Thermo-Fisher Scientific, Waltham, MA, USA). Prior to the measurement, each sample was treated to produce a flat surface through sequential grinding, cutting, and polishing. An ionization power range of 30–50 W was applied for the trace element analysis.

The crystal orientations of the 4H-SiC single crystal were evaluated using multifunction X-ray diffraction (XRD; PANalytical, Malvern, UK) and high-resolution two-dimensional (2D) XRD (D8 Discover, Bruker, MA, USA) at the Korea Basic Science Institute (KBSI, Daegu, Republic of Korea). To acquire the diffraction planes for the major and minor XRD peaks, XRD spectra were measured over the θ–2θ range using multifunction XRD with Cu Kα radiation at 30 mA and 40 kV.

The structural phase and preferential orientation of all samples were performed using a Raman spectrometer (Renishaw, Wotton-under-Edge, UK) with a 514 nm laser as the excitation source. The Raman spectra detection was acquired over the wavenumber range of 200–2000 cm^{-1} with a four-stage Peltier-cooled CCD detector (UV-Vis-NIR range). The objective lens of the microscope (DM500, LEICA, Wetzlar, Germany) had a magnification of 50$\times$, and the exposure time for accumulation was 5 s. The power intensity of the laser beam was 5.0 $\pm$ 0.1 mW in the exposure time for accumulation, which was 5 s.

The photoluminescence (PL) spectra were obtained at both 298 K and 50 K using a PL spectrometer (LabRAM HR Evolution, Horiba, Kyoto, Japan) equipped with a He–Cd laser as the excitation source (wavelength = 325 nm, power = 0.15–15 mW). The power density of the used laser ranged from 0.023 to 23.6 kW cm^{-2} for measurement. The laser was focused on the sample using a 50$\times$ objective lens.

To quantitatively monitor the elemental depth distribution in the 4H-SiC single crystal, magnetic sector secondary-ion mass spectrometry (magnetic sector-SIMS; IMS-6f, Cameca, France) was employed. For the magnetic sector-SIMS measurements, the $^{12}C^{14}N^-$ secondary ions were acquired using a cesium ion (15 kV), while the $^{11}B^+$ and $^{27}Al^+$ secondary ions were acquired using oxygen ions (7.5 kV) at a high mass resolution. Oxygen ($^{32}O_2{}^+$, 10 keV) was applied as the sputtering ion source for the detection of the secondary ions. The primary ion was rastered over an area of 200 × 200 μm^2, and the secondary ion signal was recorded from the central part of this area (~60 μm diameter). After the magnetic sector-SIMS depth profiling was completed, the formed crater depth was measured using a stylus profiler. A constant erosion rate was assumed for the conversion of the sputtering time to the corresponding depth.

To analyze the component distributions on the micro-pipe defect surfaces of the Case I and Case II 4H-SiC crystal samples, time-of-flight secondary ion mass spectroscopy (ToF-SIMS; IonTOF 5, ION-TOF, Münster, Germany) was used to measure both the positive and negative-ion modes under high-current bunching conditions. The pressure in the analysis chamber was maintained at a very low vacuum level below 1×10^{-9} Torr. The voltage and $Bi_1{}^+$ ion current were 30 kV and 1 pA, respectively. The negative-ion and positive-ion mass spectra were acquired from a 500 × 500 μm^2 area using a $Bi_1{}^+$ (1 pA) primary ion beam operating at 30 keV. The mass resolution of the measured spectra was measured under high-resolution conditions of 8000 or more at a mass-to-charge ratio (m/z) of ^{29}Si. To prevent surface charging during the ToF-SIMS measurements, a flood gun of low-energy electrons was used.

To further analyze the component distributions on the micro-pipe defect surfaces, ToF-SIMS images were recorded using the burst alignment mode, with $Bi_3{}^+$ primary ions operating at a voltage and ion current of 30 keV and 0.6 pA, respectively. The ToF-SIMS images were acquired over an area range of 500 × 500 μm^2 in both the negative-ion and positive-ion modes.

In addition, to investigate the vertical distributions of the components in the depth direction for the Case I and Case II samples, the sputter raster was set at 150 × 150 mm^2, and the secondary ions were detected using ToF-SIMS with a depth profile mode based on an area of 50 × 50 mm^2 centered within the sputtered region. A pulsed electron flood source was employed for charge compensation.

For the negatively charged secondary ions, the depth profile was measured using Cs^+—sputtering ions operating at a voltage and ion current of 3 keV and 37 nA, respectively. The depth profile of the positively charged secondary ions was measured in the non-interlaced analysis mode using $O_2{}^+$ primary ions at a voltage and ion beam current of 2 keV and 310 nA, respectively.

3. Results

Figure 5a,b shows the XRD patterns and X-ray rocking curves of the Case I and Case II 4H-SiC single crystals in the (0004) direction, wherein it can be seen in Figure 4a that all samples prepared by PVD were preferentially grown along strong (004) planes. In this figure, the XRD peaks observed at 35.6°, 38.1°, 43.2°, 49.7°, 57.2°, 65.7°, and 74.9° were attributed to the (004), (012), (013), (014), (015), (016), and (017) planes of the 4H-SiC phase, respectively [15,16,19]. All peaks were in good agreement with ICSD card 98-016-4971. In addition, the full width at half-maximum (FWHM) values for the two (0004) X-ray rocking curves exhibit similar values below 0.01°, thereby indicating that both 4H-SiC specimens are highly crystalline.

The Raman spectra of the grown 4H-SiC crystals are shown in Figure 6a,b, wherein the two characteristic peaks of 4H-SiC, corresponding to the transverse optical (TO) phonon, were detected at ~778 and 797 cm^{-1} [5,20–24]. For the Case I (bright) crystals, the TO peak observed (FTO, E_2) at 781 cm^{-1} in the perpendicular orientation was split into a doublet at 777 and 783 cm^{-1}, indicating that the 4H-SiC structure of the single crystal did not change, but the preferred orientation of the (0001) plane changed [24,25]. In Case II (dark), the TO peak (FTO, E_2) was observed at 781 cm^{-1} in the parallel orientation, and no splitting was evident. In the early stages of growth (Case I), the FTO peak intensity initially weakened prior to strengthening again in the final growth stage (Case II) until the crystal's growth was complete. The Raman spectra presented in Figure 6b for cases I and II were recorded with the incident laser oriented parallel to the c-axis.

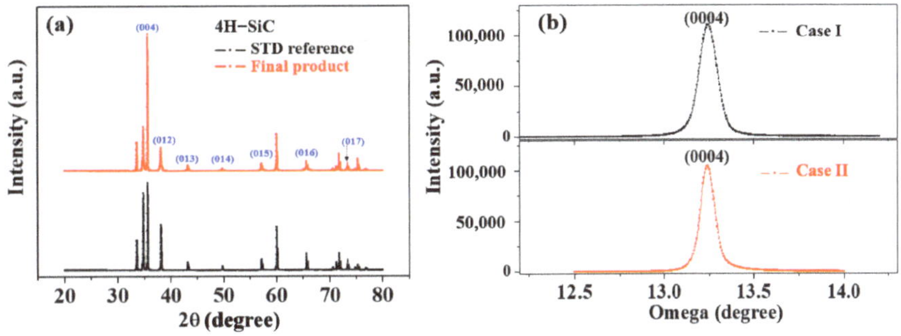

Figure 5. (**a**) XRD patterns and (**b**) X-ray rocking curves of Case I and Case II 4H-SiC single crystals in the (0004) direction.

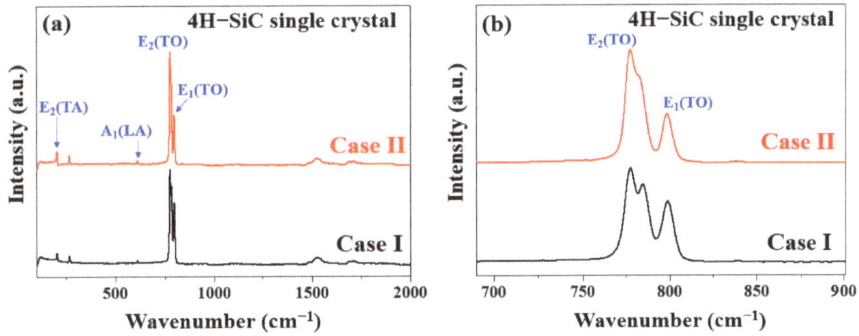

Figure 6. Raman spectra of the two 4H-SiC single crystals (cases I and II) measured under (**a**) wide and (**b**) narrow scanning conditions for the TO peaks.

Based on the obtained experimental results, it was apparent that all samples maintained the 4H-SiC structure [24]. For the Case II (dark) specimen, the folded TO peak (FTO, E_2) recorded in the perpendicular orientation at 781 cm^{-1} was split into a doublet at 777 and 783 cm^{-1} [24]. Considering the displacement of atoms from their original positions in the lattice, it was evident that this splitting occurred perpendicular to the change in the rotational c-axis of the defect [24]. Table 4 summarizes the detailed assignment and origin of the Raman peaks with respect to three different sample conditions. This splitting occurred perpendicular to the change in the rotational defect to the c-axis, considering the displacement of atoms from their original positions in the lattice [24].

Table 4. Peak center positions of the Raman peaks of cases I and II relative to the 4H-standard reference.

Sample Identification	Color	FTO, Transversal (Planar) Optic, E_2	
		∥ (cm^{-1})	⊥ (cm^{-1})
Case I	Bright yellow	N.D.	777, 783
Case II	Dark brown	781	N.D.

Figure 7a shows the PL emission spectra measured at room temperature (298 K) with a 325 excitation source for sample cases I and II. For all samples, the obtained PL spectra showed a wide non-gaussian symmetric peak at 533 nm. This PL peak corresponds to nitrogen–boron (N–B), which originated from the N–B donor–acceptor pair (DAP) emission [21,22]. The luminescence emissions displayed in these spectra are directly correlated with energy-level transitions of semiconductors. Thus, the recombination of DAPs in the 4H-SiC crystals containing an indirect bandgap leads to the formation of free excitons and phonons. This type of recombination introduces a complex donor-acceptor recombination mechanism, indicating the potential of the impurity concentration to impart a critical influence on the luminescence emissions [26–28].

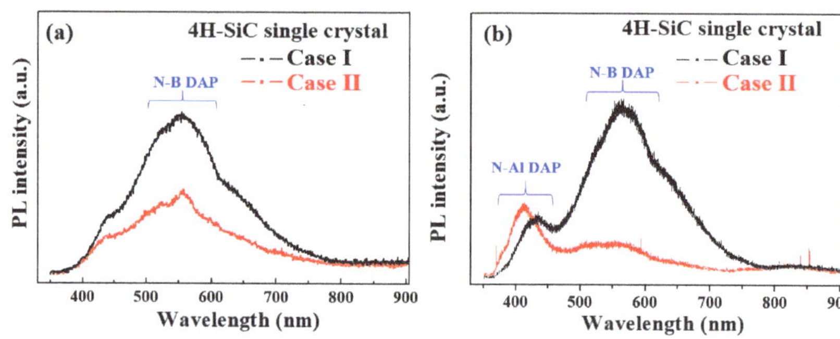

Figure 7. PL spectra of the 4H-SiC single crystals (cases I and II) measured (**a**) at room temperature (298 K) and (**b**) at a low temperature of 50 K in the wavelength range of 300–900 nm.

Figure 7b shows the PL spectra measured at a low temperature of 50 K for specimen cases I and II. In all samples, the nitrogen–aluminum (N–Al) and N–B DAP emissions were also detected and were confirmed at 420 and 580 nm, respectively [26–28], with the highest peak intensity being observed for the N–Al DAP emission of Case II. In addition, the N–B DAP emission peak intensity was high for Case I, suggesting an increase in the N–B DAP density at low temperatures [26–28]. Moreover, the peak intensity of the N–B DAP emission increased with an increasing nitrogen concentration. As shown in Figure 7b, Case I (bright) exhibits a weak PL peak at 370 nm in the low-temperature spectrum, which was attributed to the nitrogen emission caused by nitrogen-bounce excitation [26–28]. In contrast, the N–Al DAP emission was observed at 420 nm in Case II due to the effect of a visible emission at 533 nm that was caused by luminescence quenching [26–28].

To better clarify the correlation between the impurities and the PL behavior, magnetic sector-SIMS depth profiling was performed to identify the trace impurity concentrations of B, Al, and N with two different samples of cases (a) I and (b) II. As outlined in Figure 8 and Table 5, the concentration of each impurity was converted into atoms cm^{-3}, and the donor–acceptor recombination ratios (RDAs) of C_{D-A} and $2C_B/(C_N - C_B)$ were calculated using Equation (1) [15,21,22], wherein the B, Al, and N concentrations are defined as C_B, C_{Al}, and C_N, respectively. The parameter C_{D-A} was calculated as a function of $C_B - (C_{Al} - C_N)$ [16] to give the values listed in Table 5.

$$C_{D-A} = \ln C_B - (\ln C_{Al} - \ln C_N), \quad (1)$$

In Case II (dark), the increase in the PL intensity of the N–B DAP originates from an increase in the RDA of C_{D-A} [15]. Considering that the DAP recombination rate is an efficient measure of the emission luminescence [29–32], and the RDA is proportional to the donor concentration (C_D) and the acceptor concentration (C_A), the correlation between the C_D, C_A, and PL properties can be evaluated, as described previously [21]. The increased N–B DAP peak intensity for Case II was, therefore, attributed to an increase in C_N. Aukerman and Millea's model [30–32] suggests that the correlation between DAP recombination and concentration causes the PL intensity to increase with an increase in the difference between C_N and C_B; this is known as the N–B concentration gap. More specifically, when the N–B concentration gap is larger than C_B, saturation occurs, whereas when the N–B concentration gap is more than twice C_B, the PL emission intensity decreases due to the presence of non-radiative defects (non-emission), as evidenced in Case II [29]. For Case II, at a high Al concentration (C_{Al}), the N–B and N–Al DAP emissions at 420 nm became weaker [15]. As shown in Figure 8 and Table 5, for Case I, the concentrations of B, N, and Al were calculated to be 1.29×10^{17}, 3.49×10^{19}, and 5.49×10^{14} atoms cm^{-3} respectively, whereas in Case II, the corresponding concentrations were 1.99×10^{16}, 4.04×10^{19}, and 1.05×10^{16} atoms cm^{-3}, respectively. From these results, it is clear that the concentrations of N and Al were higher in Case II than in Case I. The $2C_B/(C_N - C_B)$ value was calculated to compare the correlation between the N–B concentration and PL [15,29].

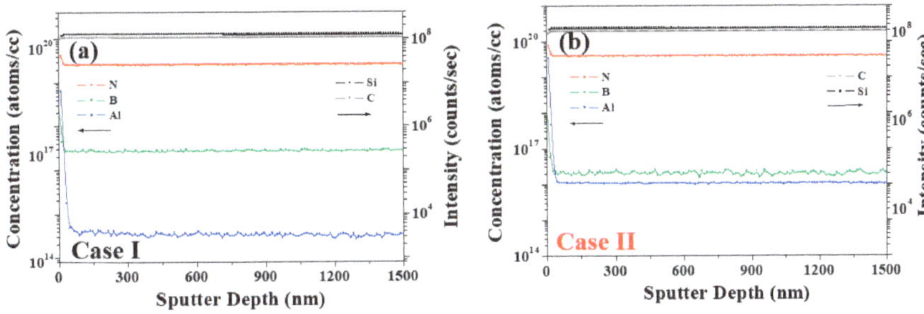

Figure 8. Depth profiling results obtained by using a magnetic sector-SIMS with two different samples of cases (**a**) I and (**b**) II.

Table 5. Comparison of the calculated elemental concentrations of B, N, and Al from the depth profiling obtained by magnetic sector-SIMS depth profiling with two different samples of cases I and II.

Concentration	Sample Conditions	
	Case I (S1, Bright)	Case II (S2, Dark)
C_N (atoms cm^{-3})	3.49×10^{19}	4.04×10^{19}
C_B (atoms cm^{-3})	1.29×10^{17}	1.99×10^{16}
C_{Al} (atoms cm^{-3})	5.49×10^{14}	1.05×10^{16}
Ln C_N	45	45
Ln C_B	39	38
Ln C_{Al}	34	37
RDA of C_{D-A} (arb. units)	50	46
The ratio of $2C_B/(C_N - C_B)$	0.01	0.001

For all samples, the peak intensity of the N–B emission at 580 nm decreased in the PL spectra, as shown in Figure 7a,b. It indicates that the N impurity has been substituted with carbon atoms within the SiC lattice; that is, these substitutions trapped the impurity levels within the bandgap of the material, and then, the impurity levels act as trap states for charge carriers (electrons and holes) and make it difficult for them to transition between the impurity levels to a non-radiative emission of 580 nm owing to the electron–hole pair recombination, resulting in discoloration and luminescence quenching.

For the Case II sample, the calculated value of $2C_B/(C_N - C_B)$ was 0.001 less than 0.01, meaning a weak visible emission at a wavelength of 580 nm. The presence of Al and N impurities in the hexagonal lattice of the SiC results in a blue emission peak that is derived from N–B DAP emission quenching. This emission peak means that the impurity concentrations of N, B, and Al have an influence on N–B DAP emission quenching in 4H-SiC [15].

To investigate the linear-type micro-pipe defects of the 4H-SiC surface, the different surface component distributions were analyzed using ToF-SIMS for the Case I 4H-SiC single crystal. Figure 9a,b shows the ToF-SIMS images obtained for the micro-pipe region of the crystal using the ToF-SIMS fast-imaging mode. For Case I, strong Na^+, Al^+, and K^+ peaks were detected in the micro-pipe defects using the positive-ion mode (Figures S1 and S2), whereas Cl^- was detected with high intensity in the negative-ion mode (Figures S3 and S4). It was therefore confirmed that the impurities of Na, Al, K, and Cl were highly distributed in the micro-pipes defect region of Case I's surface. However, no micro-pipe defects were observed on Case II's surface due to the expected distribution of impurities throughout the SiC species. It was considered that this occurred via diffusion at the high-temperature conditions employed during the final stage of crystal growth (Figure S5).

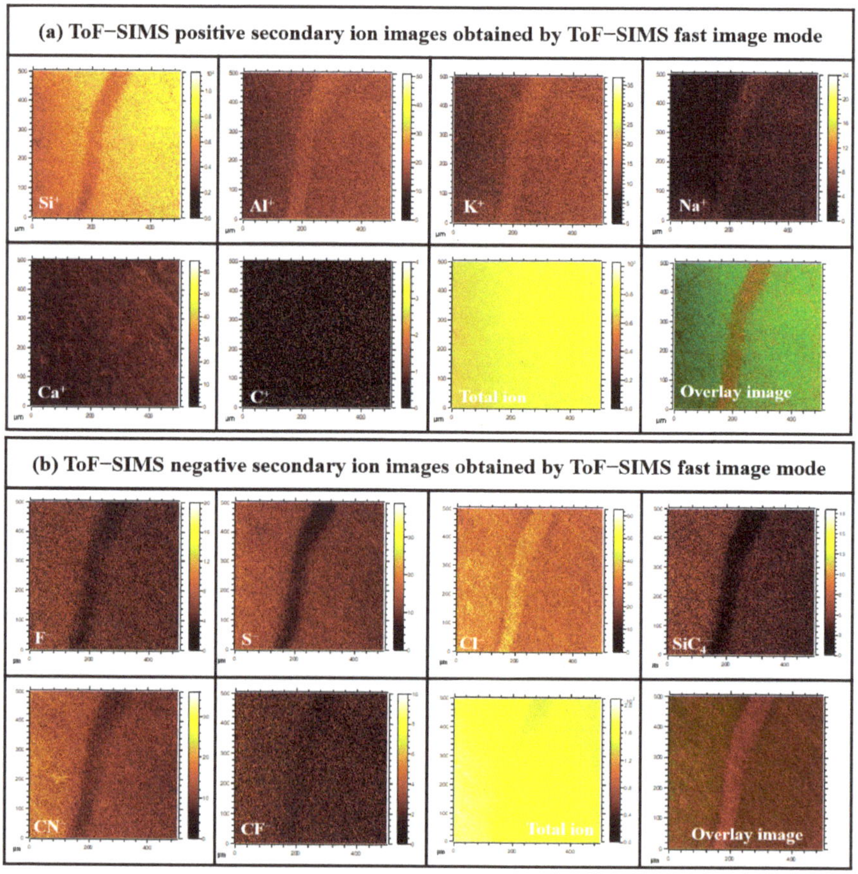

Figure 9. (**a**) Positive and (**b**) negative secondary ion ToF-SIMS images recorded in the fast image mode for the Case I (bright) sample.

To clarify the different spatial distributions in the depth direction, ToF-SIMS depth profiling was performed for Case I and Case II's 4H-SiC single crystal specimens, as presented in Figure 10a,b. More specifically, from Figure 10a, it can be seen that the impurities of Case I were distributed only on the SiC surface, compared to Case II (Figures S1 and S2). These results are consistent with the fast image results presented in Figure 9. Thus, the Na, K, Al, and Ca impurities penetrated SiC and were distributed throughout the structure for Case II, as shown in Figure 10b.

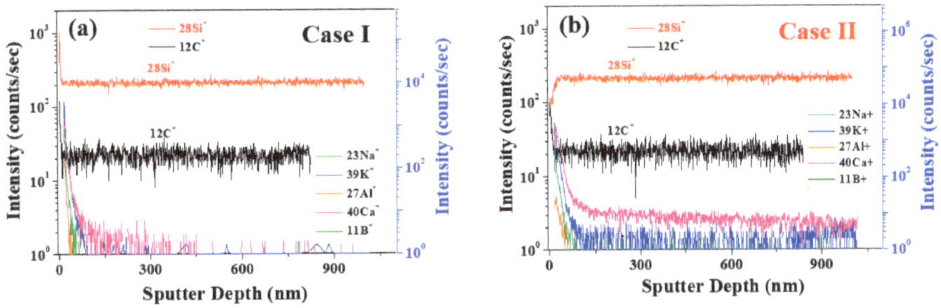

Figure 10. ToF-SIMS positive-ion depth profiling of 4H-SiC single crystal specimens with two different samples of cases (**a**) I and (**b**) II.

Figure 11a,c shows the ToF-SIMS negative-ion depth profiling results of 4H-SiC single crystal specimens with respect to the Cases I and II. More specifically, as shown in Figure 11c, the highest SiC_2^- content was detected in Case I, indicating that during this initial stage of crystal growth, the specimen is richer in carbon than during the latter stages (i.e., Case II). In Figure 11d, it can be seen that this higher carbon content can be attributed to an increase in SiC_2. It was therefore considered that this carbon species may be associated with the micropipe defects observed in the Case I specimen. Indeed, during the PVT process, the 3C-SiC precursor powder was sublimated to generate vaporized species, such as Si, Si_2C, and SiC_2, which exist in a silicon-rich and carbon excess, as indicated in Equation (2) [33–35]:

$$SiC\ (s) \rightarrow Si\ (g) + Si_2C\ (g) + SiC_2\ (g) + C\ (s) \quad (2)$$

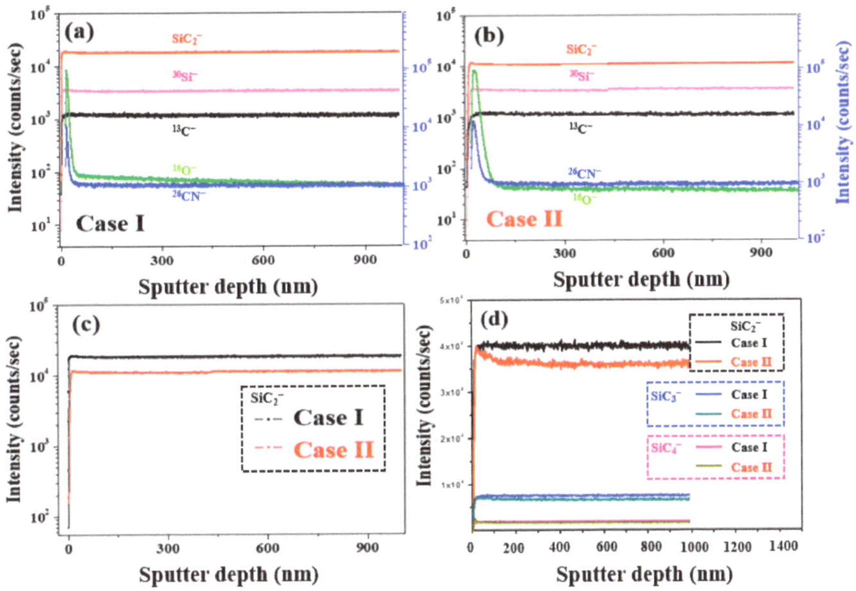

Figure 11. ToF-SIMS negative-ion depth profiling for the 4H-SiC single crystal conditions: (**a**) Case I, (**b**) Case II, and (**c**) SiC_2^-. (**d**) Comparison of the SiC_2^-, SiC_3^-, and SiC_4^- ToF-SIMS depth profiling results.

Since these vapor species are initially desorbed during the growth process, the surface of the grown SiC ingot becomes carbon-rich, as shown in Figure 11. At this time, when a metal cation with a large atomic radius, such as Ca, comes into contact with the C or

Si-rich species, impurity substitution in the SiC crystal lattice can lead to the generation of stress or discoloration [33–35]. We suggested a method to reduce impurity substitution. Firstly, in the case of the metal impurities, the metal soluble acid salts (MCl_3 and MNO_3) in the 3C-SiC powder were purified through a catalytic reaction using hydrochloric acid and esterification ($CH_3ONO + H_2O$) [15,16]. Secondly, in the case of the effect of the C/Si ratio related to the N impurity, when the C/Si ratio increases, the N impurities incorporation into a 4H-SiC crystal can be reduced by site competition rules [35]. Additionally, to maintain the C-rich conditions, the growth surface of the 4H-SiC seed was placed on the C face, in which the N molecule was very weakly adsorbed.

4. Conclusions

This work systematically investigated the main factors of discoloration switching for 4H-SiC single crystals grown using the physical vapor transport (PVT) technique. For this purpose, time-of-flight secondary ion mass spectrometry (ToF-SIMS) and micro-Raman analyses were employed. Considering the crystal structure and the distribution of trace elements, the origin of the color changes between the early and late stages of crystal growth. The ToF-SIMS results showed that in the early stages (Case I), the bright color was attributed to some formation of micro-pipe linear defects on 4H-SiC's single crystal surface during crystal growth. In the latter stages (Case II), a dark color was observed, and no micro-pipe defects were found. These differences were attributed to the abundance of SiC_2 in the initial stage and changes in the C/Si ratio as the crystal growth proceeded. Furthermore, the disappearance of micro-pipe defects in the final growth stage was likely due to the penetration and dispersion of any metal impurities inside the SiC matrix under the high-temperature conditions employed during this stage. The discoloration defects, therefore, originated from the presence of trace impurities and induced changes in the preferred orientation of 4H-SiC. Moreover, it was considered that the observed changes in the visible green light luminescence were affected by luminescence quenching during the latter growth stages. Overall, this work provides guidelines for evaluating the defect characteristics of PVT-grown 4H-SiC single crystals, with the aim of improving crystal homogeneity.

Supplementary Materials: The following supporting information can be downloaded at: https://www.mdpi.com/article/10.3390/ma17051005/s1, Figure S1: ToF-SIMS positive secondary ion spectra recorded over two different regions: (a) ROI-1 and (b) ROI-2. The spectra were acquired from the ToF-SIMS positive secondary ion images obtained using the ToF-SIMS fast imaging mode for Case I (bright); Figure S2: Comparison of the normalized ToF-SIMS peak intensities of the ToF-SIMS positive secondary ion spectra recorded over two different regions (ROI-1 and ROI-2) and acquired from ToF-SIMS positive secondary ion image obtained using the ToF-SIMS fast imaging mode for Case I (bright); Figure S3: ToF-SIMS negative secondary ion spectra recorded over two different regions: (a) ROI-1 and (b) ROI-2. The spectra were acquired from the ToF-SIMS positive secondary ion images obtained using the ToF-SIMS fast imaging mode for Case I (bright); Figure S4: Comparison of the normalized ToF-SIMS peak intensities of the ToF-SIMS negative secondary ion spectra recorded over two different regions (ROI-1 and ROI-2) and acquired from ToF-SIMS positive secondary ion image obtained using the ToF-SIMS fast imaging mode for Case I (bright); Figure S5: ToF-SIMS (a) positive and (b) negative secondary ion images obtained using the ToF-SIMS fast imaging mode for Case II (dark) in the obtained 4H-SiC single crystal.

Author Contributions: S.-K.K., Y.L. and E.Y.J. conceptualized the study. S.-K.K., H.K., Y.L. and E.Y.J. participated in the investigation. S.-K.K., T.E.H. and E.Y.J. provided the analytical tools. S.-K.K., H.K., H.S.K., T.E.H., Y.L. and E.Y.J. wrote the original draft, wrote the review, and edited it. All authors have read and agreed to the published version of the manuscript.

Funding: Korean Institute of Ceramic Engineering and Technology Commercialization Support Project, Project code no. KCS22003-0-01.

Institutional Review Board Statement: Not applicable.

Informed Consent Statement: Not applicable.

Data Availability Statement: Data are contained within the article and Supplementary Materials.

Conflicts of Interest: The authors declare no conflicts of interest.

References

1. Deng, L.; Wang, X.; Hua, X.; Lu, S.; Wang, J.; Wang, H.; Wang, B. Purification of β-SiC powders by heat treatment in vacuum. *Adv. Compos. Hybrid Mater.* **2021**, *5*, 431–437. [CrossRef]
2. Lu, W.; Tarekegne, A.T.; Ou, Y.; Kamiyama, S.; Ou, H. Temperature-dependent photoluminescence properties of porous fluorescent SiC. *Sci. Rep.* **2019**, *9*, 16333. [CrossRef]
3. Mas'udah, K.W.; Diantoro, M.; Fuad, A. Synthesis and structural analysis of silicon carbide from silica rice husk and activated carbon using solid-state reaction. *J. Phys. Conf. Ser.* **2018**, *1093*, 012033. [CrossRef]
4. Kukushkin, S. Special Issue: Silicon Carbide: From Fundamentals to Applications. *Materials* **2021**, *14*, 1081. [CrossRef] [PubMed]
5. Zhao, L.; Zhang, J.; Pfetzing, J.; Alam, M.; Hartmaier, A. Depth-sensing ductile and brittle deformation in 3C-SiC under Berkovich nanoindentation. *Mater. Des.* **2021**, *197*, 109223. [CrossRef]
6. Via, F.L.; Severino, A.; Anzalone, R.; Bongiorno, C.; Litrico, G.; Mauceri, M.; Schoeler, M.; Schuh, P.; Wellmann, P. From thin film to bulk 3C-SiC growth: Understanding the mechanism of defects reduction. *Mater. Sci. Semicond. Process.* **2018**, *78*, 57–68. [CrossRef]
7. Huseynov, E.M.; Naghiyev, T.G. Study of thermal parameters of nanocrystalline silicon carbide (3C-SiC) using DSC spectroscopy. *Appl. Phys. A* **2021**, *127*, 267. [CrossRef]
8. Gao, P.; Xin, J.; Liu, X.; Zheng, Y.; Shi, E. Control of 4H polytype of SiC crystals by moving up the crucible to adjust the temperature field of the growth interface. *CrystEngComm* **2019**, *21*, 6964–6968. [CrossRef]
9. Shin, D.-G.; Kim, B.-S.; Son, H.-R.; Kim, M.-S. Study on the growth of 4H-SiC single crystal with high purity SiC fine powder. *J. Korean Cryst. Growth Cryst. Technol.* **2019**, *29*, 383–388.
10. Arora, A.; Pandey, A.; Patel, A.; Dalal, S.; Yadav, B.S.; Goyal, A.; Raman, R.; Thakur, O.P.; Tyagi, R. Polytype switching identification in 4H-SiC single crystal grown by PVT. *J. Mater. Sci. Mater. Electron.* **2020**, *31*, 16343–16351. [CrossRef]
11. Choi, H.S.; Jeong, S.-M. Color spectroscopy for energy-level analysis of single-crystal silicon carbide. *Cryst. Growth Des.* **2022**, *22*, 2117–2123. [CrossRef]
12. Wejrzanowski, T.; Tymicki, E.; Plocinski, T.; Bucki, J.J.; Tan, T.L. Design of SiC-doped piezoresistive pressure sensor for high-temperature applications. *Sensors* **2021**, *21*, 6066. [CrossRef] [PubMed]
13. Ohtani, N.; Fujimoto, T.; Katsuno, M.; Aigo, T.; Yashiro, H. Growth of large high-quality SiC single crystals. *J. Cryst. Growth* **2002**, *237–239*, 1180–1186. [CrossRef]
14. Zhang, S.; Fu, H.; Li, T.; Fan, G.; Zhao, L. Study of Effect of coil movement on growth conditions of SiC crystal. *Materials* **2023**, *16*, 281. [CrossRef] [PubMed]
15. Kim, S.-K.; Jung, E.Y.; Lee, M.-H. Defect-induced luminescence quenching of 4H-SiC single crystal grown by PVT method through a control of incorporated impurity concentration. *Compounds* **2022**, *2*, 68–79. [CrossRef]
16. Kim, S.-K.; Jung, E.Y.; Lee, M.-H. Defect-purity evaluation of highly oriented cubic silicon carbide using micro-Raman spectroscopy. *Trans. Electr. Electron. Mater.* **2022**, *23*, 318–325. [CrossRef]
17. Ugraskan, V.; Isik, B.; Yazici, O.; Cakar, F. Surface characterization and synthesis of boron carbide and silicon carbide. *Solid State Sci.* **2021**, *118*, 106636. [CrossRef]
18. Luo, X.; Ma, W.; Zhou, Y.; Liu, D.; Yang, B.; Dai, Y. Radiation synthesis and photoluminescence property of silicon carbide nanowires via carbothermic reduction of silica. *Nanoscale Res. Lett.* **2010**, *5*, 252. [CrossRef]
19. Irfan, M.; Ajmal, M.; Mazhar, M.E.; Usmani, M.N.; Ahmad, S.; Abbas, W.; Mahmood, M.; Hussain, M. Growth and characterization of 4H-SiC by thermal evaporation method. *Dig. J. Nanomater. Biostruct.* **2019**, *14*, 243–247.
20. Liu, X.; Zhuo, S.-Y.; Gao, P.; Huang, W.; Yan, C.-F.; Shi, E.-W. Donor-acceptor-pair emission in fluorescent 4H-SiC grown by PVT method. *AIP Adv.* **2015**, *5*, 047133. [CrossRef]
21. Khashan, K.S.; Ismail, R.A.; Mahdi, R.O. Synthesis of SiC nanoparticles by SHG 532 nm Nd:YAG laser ablation of silicon in ethanol. *Appl. Phys. A* **2018**, *124*, 443. [CrossRef]
22. Aldalbahi, A.; Li, E.; Rivera, M.; Velazquez, R.; Altalhi, T.; Peng, X.; Feng, P.X. A new approach for fabrications of SiC based photodetectors. *Sci. Rep.* **2016**, *6*, 23457. [CrossRef] [PubMed]
23. Bauer, M.; Gigler, A.M.; Huber, A.J.; Hillenbrand, R.; Stark, R.W. Temperature-depending Raman line-shift of silicon carbide. *J. Raman Spectrosc.* **2009**, *40*, 1867–1874. [CrossRef]
24. Jia, Y.; Tan, Y.; Cheng, C.; Vázquez de Aldana, J.R.; Chen, F. Efficient lasing in continuous wave and graphene Q-switched regimes from Nd:YAG ridge waveguides produced by combination of swift heavy ion irradiation and femtosecond laser ablation. *Opt. Mater. Express* **2014**, *22*, 12900. [CrossRef] [PubMed]
25. Zhuo, S.-Y.; Liu, X.-C.; Xu, T.-X.; Yan, C.-F.; Shi, E.-W. Strong correlation between B-Al-N doping concentration fluctuation and photoluminescence effects of f-SiC. *AIP Adv.* **2018**, *8*, 075130. [CrossRef]
26. Nagasawa, F.; Takamura, M.; Sekiguchi, H.; Miyamae, Y.; Oku, Y.; Nakahara, K. Prominent luminescence of silicon-vacancy defects created in bulk silicon carbide p-n junction diodes. *Sci. Rep.* **2021**, *11*, 1497. [CrossRef]
27. Liu, L.; Liu, A.; Bai, S.; Lv, L.; Jin, P.; Ouyang, X. Radiation resistance of silicon carbide Schottky diode detectors in D-T fusion neutron detection. *Sci. Rep.* **2017**, *7*, 13376. [CrossRef]

28. Allgaier, R.S. Extension of the aukerman-willardson two-band hall coefficient analysis. *J. Appl. Phys.* **1965**, *36*, 2429–2434. [CrossRef]
29. Wei, Y.; Tarekegne, A.T.; Ou, H. Influence of negative-U centers related carrier dynamics on donor-acceptor-pair emission in fluorescent SiC. *J. Appl. Phys.* **2018**, *124*, 054901. [CrossRef]
30. Zhuo, S.-Y.; Liu, X.-C.; Huang, W.; Kong, H.-K.; Xin, J.; Shi, E.-W. Photoluminescence in fluorescent 4H-SiC single crystal adjusted by B, Al, and N ternary dopants. *Chin. Phys. B* **2019**, *28*, 017101. [CrossRef]
31. Arora, A.; Patel, A.; Yadav, B.S.; Goyal, A.; Thakur, O.P.; Garg, A.K.; Raman, R. Study on evolution of micropipes from hexagonal voids in 4H-SiC crystals by cathodoluminescence imaging. *Microsc. Microanal.* **2021**, *27*, 215–226. [CrossRef]
32. Yu, W.; Peng, G.; Jun, F.; Penggang, W.G.; Pei, L.; Li, Y. Research progress on the growth of 4H-SiC crystal by PVT method and the defect of polytype inclusions. *J. Sythetic Cryst.* **2022**, *51*, 2138–2152.
33. Yamaguchi, H.; Sakiyama, Y.; Makino, E.; Onda, S.; Matsumoto, Y. Ab Initio studies of the surface reaction of Si_2C and SiC_2 with Si on the 4H-SiC (000-1) surface. *Mater. Sci. Forum* **2006**, *527–529*, 235–238. [CrossRef]
34. Atkinson, C.M.; Guziewski, M.C.; Coleman, S.P.; Nayak, S.K.; Pamir Alpay, S.P. First principles analysis of impurities in silicon carbide grain boundaries. *Acta Mater.* **2021**, *221*, 117421. [CrossRef]
35. Ferro, G.; Chaussende, D. A new model for in situ nitrogen incorporation into 4H-SiC during epitaxy. *Sci. Rep.* **2017**, *7*, 43069. [CrossRef] [PubMed]

materials

MDPI

Article

SiC Doping Impact during Conducting AFM under Ambient Atmosphere

Christina Villeneuve-Faure [1], Abdelhaq Boumaarouf [2], Vishal Shah [3], Peter M. Gammon [3], Ulrike Lüders [2] and Rosine Coq Germanicus [2,*]

[1] LAPLACE (Laboratoire Plasma et Conversion d'Energie), Université de Toulouse, CNRS, UPS, INPT, 118 Route de Narbonne, CEDEX 9, 31062 Toulouse, France; christina.villeneuve@laplace.univ-tlse.fr

[2] CRISMAT UMR6508 (Laboratoire de Cristallographie et Sciences des Matériaux), Normandie University, Ensicaen, Unicaen, CNRS, 14000 Caen, France; abdelhaq.boumaarouf@ensicaen.fr (A.B.); ulrike.luders@ensicaen.fr (U.L.)

[3] School of Engineering, University of Warwick, Coventry CV4 7AL, UK; vishal.shah@warwick.ac.uk (V.S.); p.m.gammon@warwick.ac.uk (P.M.G.)

* Correspondence: rosine.germanicus@unicaen.fr

Abstract: The characterization of silicon carbide (SiC) by specific electrical atomic force microscopy (AFM) modes is highly appreciated for revealing its structure and properties at a nanoscale. However, during the conductive AFM (C-AFM) measurements, the strong electric field that builds up around and below the AFM conductive tip in ambient atmosphere may lead to a direct anodic oxidation of the SiC surface due to the formation of a water nanomeniscus. In this paper, the underlying effects of the anodization are experimentally investigated for SiC multilayers with different doping levels by studying gradual SiC epitaxial-doped layers with nitrogen (N) from 5×10^{17} to 10^{19} at/cm^3. The presence of the water nanomeniscus is probed by the AFM and analyzed with the force–distance curve when a negative bias is applied to the AFM tip. From the water meniscus breakup distance measured without and with polarization, the water meniscus volume is increased by a factor of three under polarization. AFM experimental results are supported by electrostatic modeling to study oxide growth. By taking into account the presence of the water nanomeniscus, the surface oxide layer and the SiC doping level, a 2D-axisymmetric finite element model is developed to calculate the electric field distribution nearby the tip contact and the current distributions at the nanocontact. The results demonstrate that the anodization occurred for the conductive regime in which the current depends strongly to the doping; its threshold value is 7×10^{18} at/cm^3 for anodization. Finally, the characterization of a classical planar SiC-MOSFET by C-AFM is examined. Results reveal the local oxidation mechanism of the SiC material at the surface of the MOSFET structure. AFM topographies after successive C-AFM measurements show that the local oxide created by anodization is located on both sides of the MOS channel; these areas are the locations of the highly n-type-doped zones. A selective wet chemical etching confirms that the oxide induced by local anodic oxidation is a SiOCH layer.

Keywords: local oxidation; atomic force microscopy; C-AFM; doping; successive AFM characterizations; doped silicon carbide; MOSFET

Citation: Villeneuve-Faure, C.; Boumaarouf, A.; Shah, V.; Gammon, P.M.; Lüders, U.; Coq Germanicus, R. SiC Doping Impact during Conducting AFM under Ambient Atmosphere. *Materials* **2023**, *16*, 5401. https://doi.org/10.3390/ma16155401

Academic Editors: Marilena Vivona and Mike Jennings

Received: 30 June 2023
Revised: 25 July 2023
Accepted: 28 July 2023
Published: 1 August 2023

1. Introduction

Recently, silicon carbide (SiC) has become a mature wide-band-gap semiconductor in microelectronics networks and especially in electric-power conversion [1] and has shown continuous high-temperature integrated circuit operation over 800 °C [2]. It has demonstrated exceptional characteristics compared to silicon, with numerous advantages in high-temperature and high-frequency applications due to its high-critical electric field and its high thermal conductivity [3]. SiC devices have already demonstrated their excellent performances in terms of energy conversion and system losses, with increased devices density [4]. SiC MOSFETs have shown their interest for power networks, especially concerning the resistance to high voltages and currents [5].

The high integration density for small scale devices, the process requirements, but also the reliability analysis require precise and local techniques to analyze devices at the SiC-die level. The details of the SiC inside the die, i.e., the achieved dopant level and other fundamental properties, as well as the interactions at its interfaces with other materials are important to be analyzed to understand the macroscopic performances of the full device, compared to its modelled attributes. Based on atomic force microscopy (AFM), several modes are becoming essential techniques to probe and examine local electrical properties at the nanoscale. These modes include modes based on conduction-like scanning capacitance microscopy (SCM) [6,7], scanning non-linear dielectric microscopy (SNDM) [8–10], conductive AFM (C-AFM) [11–15] or scanning spreading resistance microscopy (SSRM) [16–19], and a mode based on microwave interaction sMIM (scanning microwave impedance microscopy) [20,21]. Nowadays, all these modes are used extensively for their capability in terms of spatial resolution, detection, analysis of doping species, conductivity, resistivity or isolation properties. During the measurements with these modes, an electrical voltage is applied on the tip scanning the surface of the sample, with a mechanical contact between the tip and the surface. When the measurements are carried out under ambient atmosphere, i.e., air containing moisture, the formation of a water meniscus surrounding the tip occurs due to the adsorption of the humidity. This water bridge changes the local electrochemical behavior around the AFM tip and can cause a localized oxidation of the material under investigation [22]. For silicon, several extended studies have been conducted since the 1990s. At the tip–sample nano-contact, a local electrochemical cell is created (Figure 1), where the AFM silicon tip plays the role of a cathode. When a negative appropriate bias is applied to the tip, the high electrical field (in the range of 10^9–10^{10} $V{\cdot}m^{-1}$) in the electrochemical nano-cell induces the creation of O^-, OH^-, and H^- ions. The polarization applied on AFM favors water electro dissociation. With the molecular transport (ionic-diffusion mechanisms) through the water meniscus, negative ions flow towards the surface and recombine with holes (h+) [23] at the surface of the silicon. Therefore, this anodic oxidation mechanism allows for a direct oxide growth located under the AFM tip.

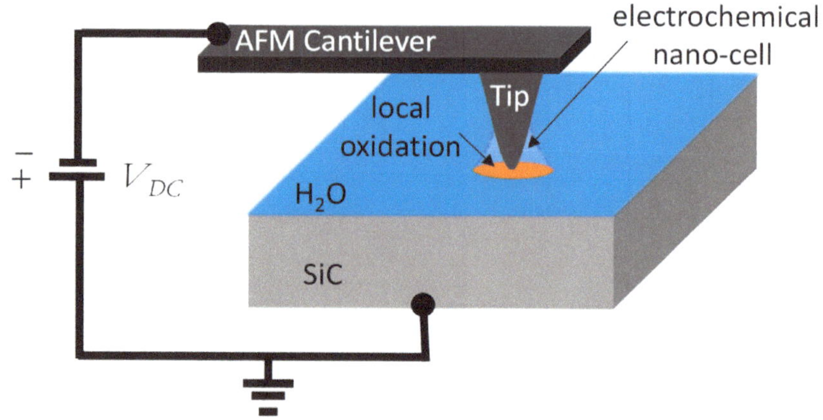

Figure 1. AFM tip in contact with the SiC with the presence of absorbed water: creation of an electrochemical nano-cell by applying a negative bias to the tip and local anodization oxidation.

This mechanism is called AFM local anodic oxidation (AFM-LAO). AFM-LAO is used to create and manage oxidized nano-structures like nano-dots or lines also called (nano-patterning). The grown oxide thickness depends on several parameters such as the silicon properties and scanned AFM parameters [24].

For SiC, authors have reported such a local oxidation by a polarized AFM tip during the scan [25,26]. As for the silicon, the main origin of this phenomenon is due to the ionization of the water meniscus presence under electric fields of at least 8×10^6 V/m [25,27,28]. The AFM-LAO oxide growth rate for SiC was studied as a function of the applied DC bias [29,30], bias pulses [30], loading force, scan rate and/or doping level for the 6H-SiC and 4H-SiC [31]. In [31], authors also show that the AFM local oxidation can be promoted by high doping concentrations; in this case, two doping levels of phosphorus-implanted 4H-SiC are experimentally compared.

The growth mechanisms during this local anodization and, more particularly, the influence of the SiC doping have not been comprehensively investigated up to now. In this context, this study aims to investigate the influence of the SiC doping level on this growth mechanism during conductive AFM scans. This study goes a step further considering two aspects: (i) the fine quantification of the local oxidation as a function of the doping level with the study of a gradual SiC epitaxial-doped layers with nitrogen (N); and (ii) the study of the local oxidation mechanism after consecutive scans of the cross-section of a power SiC MOSFET. To reach this goal, a staircase doping sample is scanned and analyzed. The experimental results provided by AFM are combined with finite element modeling, to evaluate the electric field and current distributions when the tip is in contact with the sample surface. Results are extended to a classical SiC MOSFET by studying the evolution of the surface cross-section after successive C-AFM scans.

2. Materials and Methods

2.1. Description of the SiC Staircase Sample and MOSFET Device

Two kinds of samples were used (Figure 2): epitaxial SiC layers with various doping levels, and a commercial off-the-shelf (COTS) SiC MOSFET. The epitaxial SiC (sample A) was specially designed for this study; the sample is a staircase SiC multilayer doped in situ with nitrogen (N) with dopant concentrations from 5×10^{17} to 10^{19} at/cm^3. This sample is fabricated to determine the threshold of the doping level for the LAO mechanism. Epitaxial layers were grown at Warwick on 100 mm diameter, 4° off-axis, n+ 4H-SiC wafers using an LPE ACiS M8 chemical vapor deposition (CVD) reactor. The 6 μm thick 4H-SiC homoepitaxial layers were grown using trichlorosilane (HCl_3Si, TCS) and ethylene (C_2H_4) at a C/Si ratio of 0.46 with nominal growth rates of 5 μm/hr at a temperature of 1650 °C, where N_2 was used to dope the layers. An optical view of the sample A is shown in Figure 2; note that another SiC wafer is used to avoid edge effects during the polishing state. Sample B is a classical TO-molded packaged COTS power SiC MOSFET that has a planar structure (third generation) with a vertical double implantation. The nominal voltage and current rating are 900 V and 23 A, respectively, and R_{DS} = 12 Ω. For AFM scans, a cross-sectional sample was prepared, keeping the electrical access through the TO packaging for electrical biasing during the AFM scans [32]. The axis of the cut is perpendicularly to the gate contacts. The surface of the cross-sectional sample was polished with lapping films. Decreasing granularity disks were used to obtain a very low roughness, i.e., a mirror-like surface. The schematic cross-section SiC epilayers represent the multilayer structure deposited on the 4H-SiC substrate. A confocal view (Figure 2) reveals the multigate structure of the device. A schematic of the MOSFET structure is also reported.

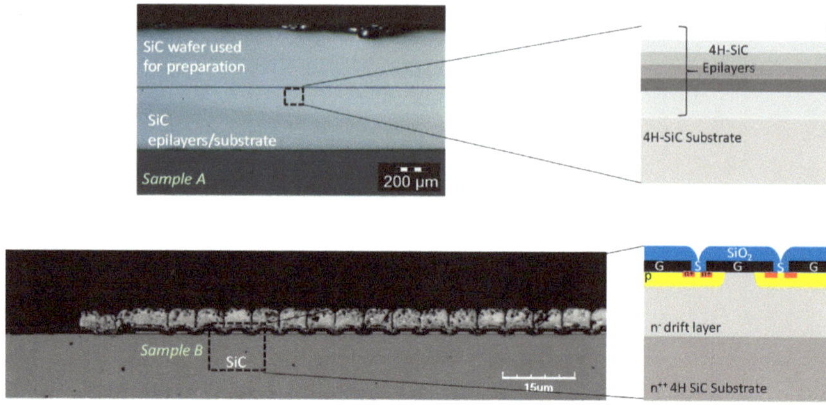

Figure 2. Studied SiC samples: Sample A, optical view of the cross-section of the epitaxial SiC multilayer deposited on the 4H-SiC substrate and Sample B, confocal microscopy view of the cross-section after the surface preparation and a schematic of the classical MOSFET structure.

2.2. Atomic Force Measurement Details

For the AFM measurements, a Dimension Icon AFM from Bruker is used. The topography and the local conductivity were simultaneously recorded in the C-AFM [7] mode. A conductive platinum–iridium (PtIr) coated tip (SCM-PIC, Bruker, Billerica, MA, USA) was used with a tip radius around 25 nm. The measured spring constant of the used cantilever was determined to be about 0.1 N/m. This low spring constant assures a good electrical contact with the oxide at the surface of the cross-sectional sample without indenting the thin oxide layer. The image resolution was 512 × 512 pixels with a slow scan rate of about 0.3 Hz during the electrical measurements. AFM acquisitions were performed under ambient conditions at room temperature with a relative humidity of 45%.

2.3. Simulations of the Electric Field and Current Distribution under the Tip–SiC Nanocontact

In ambient conditions, the electric field located around the AFM tip and the presence of adsorbed water are the origin of the local oxidation of the scanned surface [33,34]. In fact, during AFM-LAO, by applying a negative bias to the AFM tip, a local electrochemical cell was formed around the tip and could allow local anodic oxidation (Figure 1). In order to determine the electric field distribution in the nano-system, the water meniscus/oxide/SiC region, an electrostatic model, was deployed. A 2D-axisymmetric finite element model was developed with COMSOL Multiphysics. The AFM tip was computed as a classical truncated cone with 10 µm height and 14° aperture angle, ending with a semi-spherical apex (curvature radius R_c = 25 nm). In our model, the water meniscus was simulated with a height of 30 nm and a radius of 50 nm at the interface with the surface sample. The rest of the tip was supposed to be surrounded by air, modeled by a box whose dimensions are large enough to avoid edge effects [35]. The SiC layer sample is modeled by a 500 nm thick material covered by an oxide layer (representing the LAO oxide) whose thickness ranged from 0 nm to 6 nm. As the presence of carbon cannot be excluded, a SiOCH [30,36] layer was simulated as the oxide layer.

The local electric field $\vec{E}$ derives from the potential V:

$$\vec{E} = -\overrightarrow{grad}\,(V) \quad (1)$$

Moreover, to simulate the local electric field distributions $\vec{E}$, two approaches were developed. First, to investigate the influence of water bridge presence and SiOCH layer thickness on electric field distribution, the Poisson's equation was solved in the surrounding air, in the water meniscus and in the device.

$$div\left(\varepsilon\vec{E}\right) = \rho \quad (2)$$

where ρ is the charge density and ε the dielectric permittivity.

The relative dielectric permittivity of water, SiC and SiOCH are 80.0, 9.2 and 3.0, respectively.

For the classical simulations, typical boundary conditions were applied; no charge conditions (i.e., zero potential) were applied on the free boundaries of the simulation box to avoid edge effects. Moreover, the water meniscus, SiOCH and SiC were considered free of charge (i.e., ρ = 0 C cm^{-3}).

Secondly, to investigate the influence of SiC doping on the current, the following equation was solved:

$$\vec{J} = \sigma\vec{E} \quad (3)$$

with σ the conductivity of the different materials.

In the water bridge, the conductivity is considered to be constant (5×10^6 S/m) and variable for the SiC and SiOCH layers. For the SiOCH layer, a conductivity ranging from 1.7×10^{-5} S/m (Si in excess) to 1 S/m (defect or impurities) is considered to reproduce the poor insulating properties of this oxide. For the SiC layer, a conductivity ranging from 1.7×10^3 S/m (i.e., doping concentration of 10^{17} at/cm^3) to 2×10^4 S/m (i.e., doping concentration of 5×10^9 at/cm^3) was used.

3. Results and Discussion

3.1. The Tip–Sample Interaction in Ambient Conditions: Signature of the Water Meniscus

In AFM, the force–distance curves (FDCs) provide quantitative acquisitions of forces acting between the AFM tip and studied sample, at a nanometer scale. FDCs allow for the probing of the forces inherent to the tip–sample interaction: Van der Waals, electrostatic, adhesion or mechanical forces [37,38]. More particularly, when the tip retracts from the surface, the adhesion signal between the AFM tip and surface represents the superposition of forces due to the electrostatic force F_{el}, the Van der Waals force F_{VdW}, the capillary forces F_{cap} and chemical bonds or acid–base interactions F_{chem}. F_{cap} depends on the relative humidity and the tip/sample hydrophobicity, and F_{el} could modify the water meniscus shape around the tip.

In spectroscopy mode (acquisition on a unique pixel without scanning), AFM FDC were recorded. During the tip–sample approach and retract movements, by measuring the AFM cantilever deflection, the force F follows this relation:

$$F = kSd \quad (4)$$

with k the experimental spring constant of the cantilever, S the sensitivity of the AFM cantilever and d is the laser deflection induced by AFM cantilever bending.

Experimental results are shown in Figure 3: the AFM apex interacts with the surface of the SiC in ambient conditions, with and without negative bias. Approach and retract curves are reported. These experiments were done for the SiC epitaxial layer with a doping level of 4×10^{17} at/cm^3. Without polarization, far from the SiC surface, the force applied to the AFM cantilever is zero. During the approach, when the tip is close to the surface, the Van der Waals attractive force occurs. The following linear relationship between force and Z position is characteristic of the mechanical properties of the sample surface. In a second step, when the tip is retracted from the surface, the adhesion force increases the spring constant of the cantilever and a supplementary force is needed to pull off the tip from the surface. For our experiment without bias (Figure 3a), a significant adhesion force of 1.9 nN was measured. As the Van der Waals force is low (not observed in the approach curve), the adhesion force is mainly related to capillary condensation (F_{chem} is lower than F_{VdW}). When a negative DC bias voltage of −10 V is applied on the tip (sample back-side is grounded), a modification of the FDC is observed (Figure 3b). During the tip approach, a curvature of the FDC upon contact is detected. This attractive force corresponds to the long-range electrostatic force. Indeed, an attractive electrostatic force of around 0.8 nN was measured at the contact point. During the retract movement, the adhesion force needed for pull-off increases to 3 nN (compared to the curve without polarization), which corresponds to an enhancement factor close to 1.5. This AFM measurement proves that a nanomeniscus is present around the tip, leading to a supplementary, strong adhesive force [39,40]. When a negative V_{DC} is applied to the AFM probe, the generated local electric field induces the growth of the water meniscus located around the tip. The FDC modification under electric bias is related to the increase in the electrostatic force and interaction with the liquid meniscus formed by capillary condensation at the tip–sample contact around the AFM tip. Schematics of the water bridge (capillary condensation) around the conductive AFM tip are shown in Figure 3. From the water meniscus breakup distance measured without and with polarization, the water meniscus volume is increased by a factor of three under polarization [41].

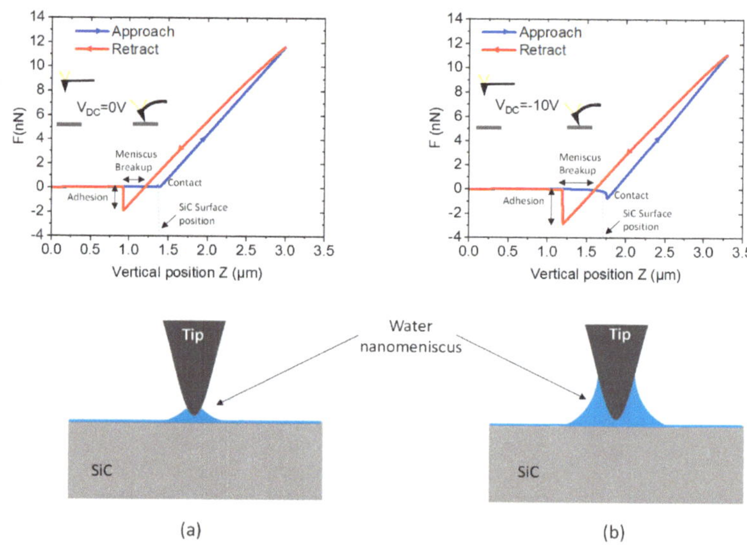

Figure 3. AFM force–distance curves for two applied DC bias: (**a**) V_{DC} = 0 V and (**b**) V_{DC} = −10 V. Approach and retract curves are with schematics of the cantilever position, and schematics of the water nanomeniscus bridging the AFM tip and the sample surfaces for the two polarizations.

The local oxidation growth mechanism of metals, described by the Mott–Cabrera model [42], is due to the existence of the water nanomeniscus with a high local electric field. In fact, the local water nanomeniscus bridges the electrical conduction mechanism and provides the anions to permit the chemical oxidation by driving hydroxyl anions towards the surface of the sample. In order to determine the electric field distribution in the near environment of the conductive AFM tip during the contact with the 4H-SiC surface, an FEM model was developed. The electric field was computed using Equations (1) and (2) and taking into account the dielectric permittivity of water, air, and the SiOCH and SiC layers. Figure 4a represents the electric field distribution when the tip is in direct contact with the SiC surface (i.e., no SiOCH layer) and without a water meniscus. As is expected in an ideal vision for nanoscale probing, the electric field is limited to the vicinity of the contact point between the AFM tip and SiC. Inside the SiC sample, an electric field E of around 8.2×10^8 V/m is obtained under the contact point and is divided by a factor 10 at a distance of 20 nm (in depth) away from this point. The presence of the water nanomeniscus (Figure 4b) fundamentally modifies the electric field distribution around the AFM tip and in the SiC layer. The electric field at the SiC surface is lowered by a factor of 10 compared to the results without a water meniscus. It is more homogeneously distributed over the interface area between the absorbed water and 4H-SiC. However, the field is now concentrated at the rim of the water meniscus and the interface with the SiC, leading to an enhanced field at a position, which is not situated under the tip, but given by the extrinsic shape of the water nanomeniscus.

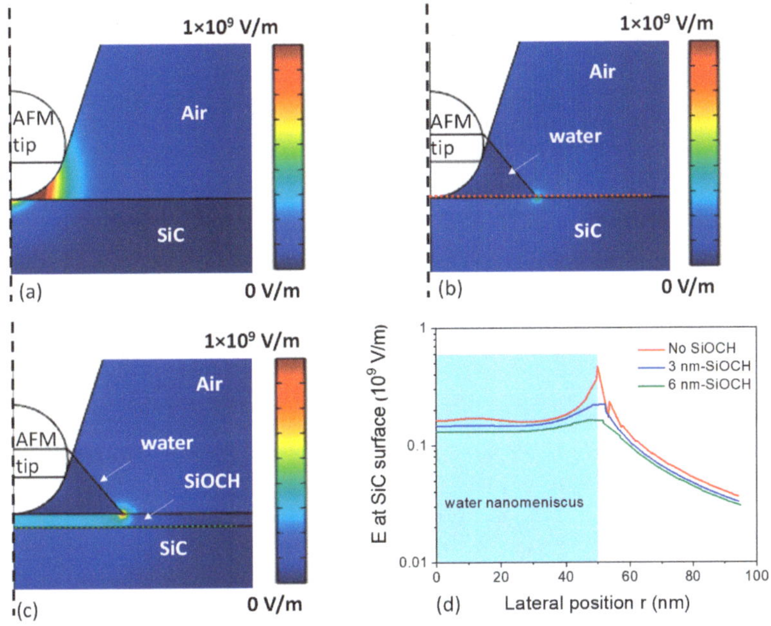

Figure 4. FEM simulation of the electric field distribution (**a**) without water, (**b**) with a water meniscus and (**c**) with a 6 nm thick oxide layer (10 V is applied on the AFM tip). (**d**) Electric field profile along SiC surface for various oxide thickness.

With the presence of an oxide layer, the electric field distributions (at the SiC surface and inside materials) are shown in Figure 4c. An enhancement of the electric field inside the oxide layer and around the edge of the AFM tip is discerned. Again, the location with the highest electric field is the point of the rim of the water meniscus at the sample surface. Profiles of the electric field at the SiC surface, or SiC/oxide interface, in the presence of a water meniscus, are represented in Figure 4d. The electric field is quasi-constant over 40 nm (i.e., water meniscus width) with a value around 1.5×10^8 V/m. The impact of the oxide thickness was also studied, and oxide thicknesses from 3 nm to 6 nm were computed. Results show that the oxide thickness has only a weak influence on the electric field distribution at SiC surface both in air as well as in the water. It is interesting to note that the presence of the SiOCH layer decreases only weakly the electric field at the SiC surface. This implies that the growth of the oxide layer during anodization by the AFM tip slowly decreases the electric field and is consequently a self-limiting growth phenomenon. Far from the meniscus, a strong decrease in the field is calculated. These results also imply that the electric field distribution will be influenced by the shape of the water meniscus (i.e., the variation of the environmental humidity) and the shape of the tip (i.e., tip ageing during contact C-AFM measurements) leading to a variety of anodization area sizes and oxide thicknesses.

3.2. Electrical Conduction at the Nanoscale

In order to evaluate the effect of the doping level of the SiC layer on the anodization, different epitaxial layers have been studied. In Figure 5, AFM topographies of the n-doped 4H-SiC sample with different doping layer levels before and after C-AFM measurement are presented. During C-AFM, a bias of −10 V applied to the conductive tip. The scan size is 5 μm × 2.5 μm (with a resolution of 512 × 256 pixels). The evolution of the SIMS profile of the nitrogen doping is superimposed on the AFM mappings as a guide to the eye. After preparation, the cross-sectional sample presents a low surface roughness. Indeed, the average surface roughness Ra and root mean square roughness Rq are equal to 0.5 nm and 0.7 nm, respectively. A surface modification was observed after C-AFM measurement in the highly doped regions (zones 4, 5 and the substrate), as can be seen from the topographical maps taken at the same location before and after the measurement (Figure 5a). Under polarization, the silicon AFM tip could be considered as a cathode and SiC as an anode. This surface modification is related to the direct oxide growth by LAO. A zoom view of 1.2 μm × 1.2 μm is reported in Figure 5b. Where the doping concentration is maximal (zone 5), we observed a change in height of 2.8 nm while an increase in roughness was observed for the regions with lower doping (zone 4 and substrate). It seems likely that the local nature of the interaction between sample and tip leads to a pixel-by-pixel anodization of the surface during the AFM scan and thus an inhomogeneous effect. Where the doping level is lower (zones 1, 2, 3 and 6 in Figure 5a,b), we did not observe a significant modification of the sample surface. This is clearest in zone 6, which becomes easy to distinguish topographically after local anodization. The AFM-LAO intensity correlates strongly and selectively with the SiC doping level as a result of local surface oxidation due to negative tip bias during C-AFM scans.

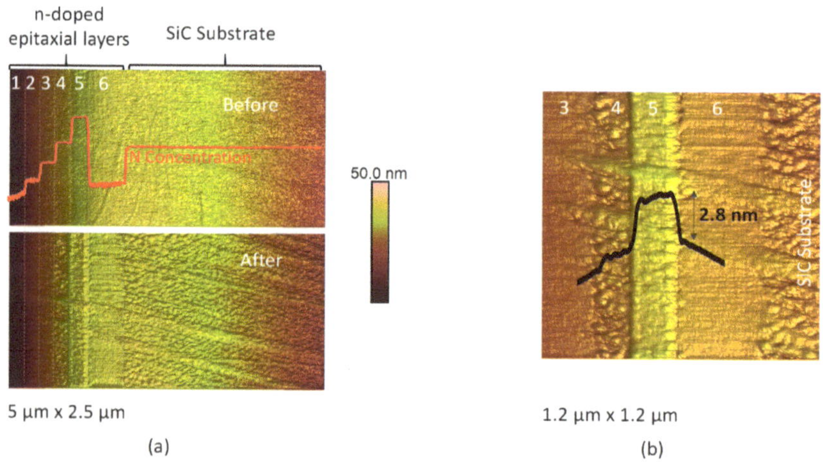

Figure 5. (**a**) AFM topographies (5 μm × 2.5 μm) before and after the AFM-LAO for the SiC-doped staircase sample; the SIMS profile is reported and (**b**) so is a zoom on the grown oxide.

Figure 6 shows the profiles of the topography along the epilayers before and after AFM-LAO. Before AFM-LAO, the form of the topography is due to the surface preparation. Compared to the topography after the C-AFM scan (after AFM-LAO), the change in height is confirmed for the layers 4, 5 and the SiC substrate. Therefore, the doping level threshold where LAO appears is is 7×10^{18} at/cm^3.

With the C-AFM, the local current can be measured in the spectroscopy mode or during the scanning measurement. Figure 7 reports the measured current during the tip approach and retract for the SiC layer with a doping level of 4×10^{17} at/cm^3. This curve is recorded during the experiment presented in Figure 3b ($V_{DC} = -10$ V). Far from the sample surface, no current flows from the tip through the back contact of the sample. After the tip's contact with the sample, at z = 1.9 μm, an increase in the current is measured. When the contact is maintained (i.e., z > 1.9 μm), a current of −2 pA is probed through the tip and the back contact. During the retract step, the collected current follows the force curve and current is collected through the water meniscus, even if the physical contact between the tip and the SiC surface is removed. This signal attests to the collection of the current through the water meniscus during this approach–retract experiment at $V_{DC} = -10$ V. As a consequence of the high electric field, the local current also participates with the LAO mechanism.

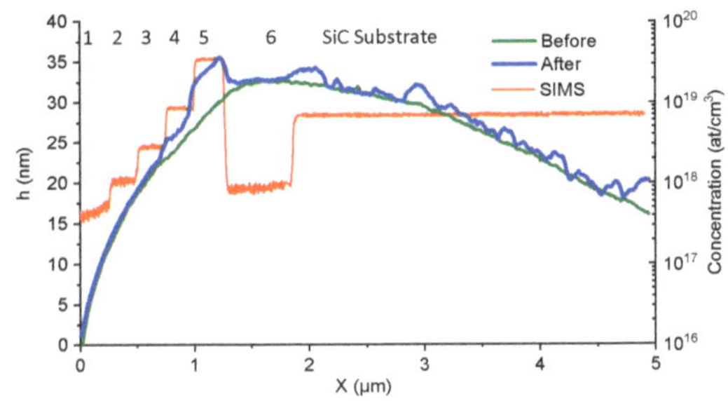

Figure 6. Profiles of the topography along the epilayers before and after AFM-LAO. The red line indicated the N concentration from SIMS technique. The doping layers are identified with numbers from 1 to 6.

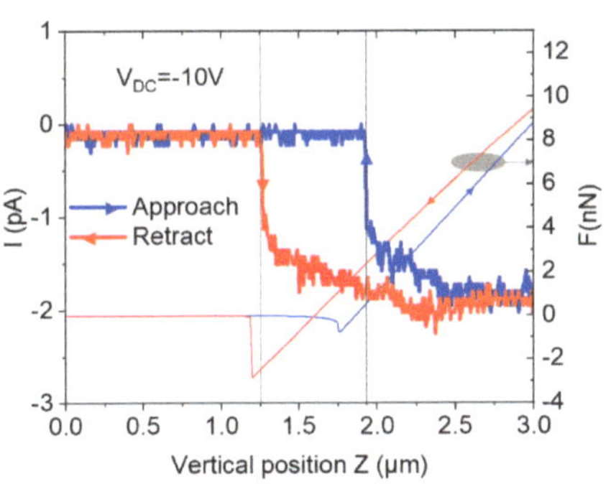

Figure 7. Evolution of the collected C-AFM current during the tip approach and retract with an applied bias voltage $V_{DC} = -10$ V. The curve forces are superimposed.

Simultaneously to the recorded topography, the local currents can also be measured. For the 4H-SiC staircase sample, the current profile along the epilayers are plotted in Figure 8. As expected, the current depends on the doping level of the SiC layer. To probe the electrical properties of the grown oxide layer, currents are collected before and after LAO. For the highest doping level (#5), the decrease in the current after LAO confirms the insulating behavior of the grown oxide.

To quantify the influence of the doping level on the anodization and on the measured current by C-AFM, the current density collected by the AFM tip is computed by FEM using Equations (1) and (3). The current density distributions calculated without oxide and with an oxide layer are shown in Figure 9a,b, respectively. The resulting currents collected by the AFM tip for different doping levels are reported in Figure 9c,d. Without the oxide layer, the current increases by a factor of 10 from the lowest to the highest doping level. Moreover, depending on this doping level, two regimes can be identified. At low doping (less than 5×10^{18} at/cm^3), the current only increases slowly. At a high-doping level (higher than 5×10^{18} at/cm^3), the current increases much more drastically. To simulate the effect of the

LAO, similar simulations are performed for the presence of oxide layers with three different conductivities (Figure 9d). The oxide presence implies a decrease in the current collected by the AFM tip with increasing SiOCH thickness. This is consistent with the observed decrease in the C-AFM current after anodization, as experimentally observed in Figure 8. Moreover, the FEM approach demonstrates that the influence of the SiOCH thickness is more pronounced for oxides with high conductivity, which indicates low-quality oxides with defects or impurities leading to bad insulating properties.

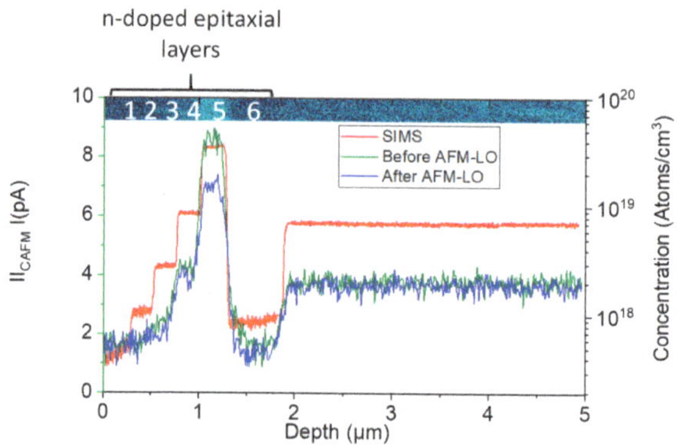

Figure 8. Profiles of the current along the epilayers before and after AFM-LAO. The red line indicated the N concentration from SIMS technique.

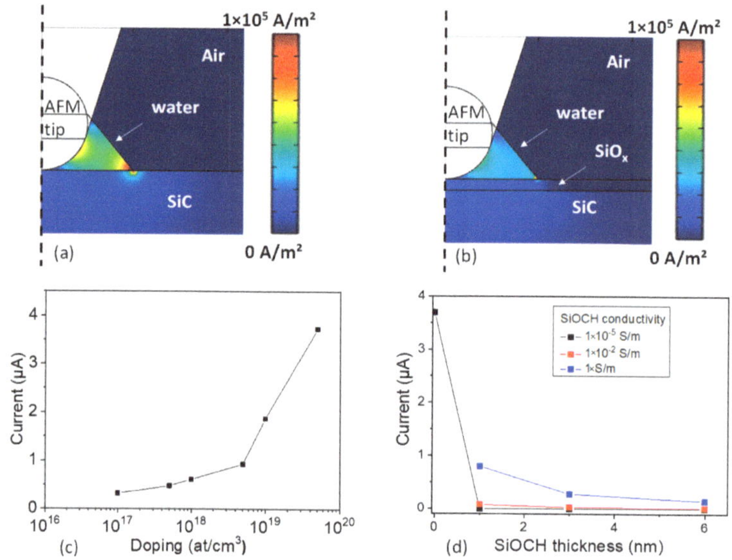

Figure 9. (**a**) FEM simulations of the surface current distribution without the oxide layer and (**b**) as function of SiC doping; −10 V is applied on AFM tip. (**c**) FEM simulation of the surface current distribution with the oxide layer and (**d**) evolution of the current collected by AFM tip as function of oxide thickness and conductivity.

A dependence of the anodization with the doping level of the SiC was shown experimentally. According to FEM, this anodization occurs for the conductive regime in which the current depends strongly to the doping level. Consequently, anodization and oxide growth are favored by high currents flowing through the tip–sample contact and high electric field applied to the tip. Indeed, a thicker oxide strongly decreases the current flowing through the SiC and the electric field distribution. Moreover, the FEM results demonstrate that the decrease in the current during C-AFM measurement is related to the presence of a poorly insulating SiOCH layer whose conductivity is around 1 S/m.

4. Results for the SiC-MOSFET

A commercial SiC-MOSFET is studied, characterized and analyzed. The topography acquired on the cross-sectional sample is reported on Figure 10a. The topography reveals the SiC MOSFET structure at the die level. The gate (G) and the source (S) contact can be recognized. Moreover, around the gate, SiO_2, inter-metal dielectric layers and aluminum contact for the source can be identified. The mapping of the AFM deflection signal error is also reported in Figure 10b, allowing to better appreciate the structure of the device.

Figure 10. AFM topographies of a commercial SiC-MOSFET (**a**) topography and (**b**) deflection error signal.

In Figure 11, the surface topography evolution for successive C-AFM acquisitions is presented. For all measurements, a negative voltage of −10 V is applied to the AFM tip during scanning. By comparing with the acquisition before applying the negative voltage, the AFM acquisition of the topography reveals new areas with a higher topographic contrast. These news areas are the oxide created during the local oxidation of the SiC. Very interestingly, the areas are located on both sides of the MOS channel, at the location of the highly n-type-doped layers (Figure 11a). For the less-doped regions (for example, the n-drift layer), no oxide formation was detected for the used experimental parameters, as expected from the results of the previous section showing the strong dependence on the SiC doping level. For the second (Figure 10b) and the third (Figure 11c) scans with negative bias, the zones showing a local oxidation stay the same, but with an enhancement of the LAO effect. During successive scans, for the LAO oxide growth, two mechanisms are involved: (i) the decreasing of the current density in the water meniscus which decreases OH^- transport to the anode and (ii) a diffusion barrier for OH^- to reach the SiC surface. This implies that oxide growth is an auto-limiting effect.

In addition, to reveal the nature of the grown oxide by LAO, a selective wet chemical etching is performed on the sample. A highly concentrated solution of potassium hydroxide (KOH, 2M) is used to dissolve SiO_2. This choice of the etching solution was made so as to slowly etch the pure SiO_2 layers in order to avoid as much as possible a surface degradation of the sample, and at the same time, preserve the pure SiC zones [43], which will act as reference zones for the evaluation of the height changes. The sample is immerged for around 10 min in the solution at room temperature. Topographies are recorded before and after the etching. For this experiment, in order to improve the resolution of the topography scan, the Scan Asyst mode is used with a AFM tip with a radius of 2 nm. Figure 12a,b show the impact of the KOH etching on the cross-section after LAO, in the areas cobining the anodic oxide layers on the highly doped SiC situated just below the SiO_2 gate oxide. The comparision of the topography profiles (Figure 12c) along the axes indicated in the respective figures before and after etching highligths a small effect on the SiO_2 gate oxide, which is more important than on the oxide layer grown by LAO. The fact that the LAO-grown oxide is etched more slowly than pure SiO_2 is a strong indication that the oxide layer grown by LAO is mainly a SiOCH layer.

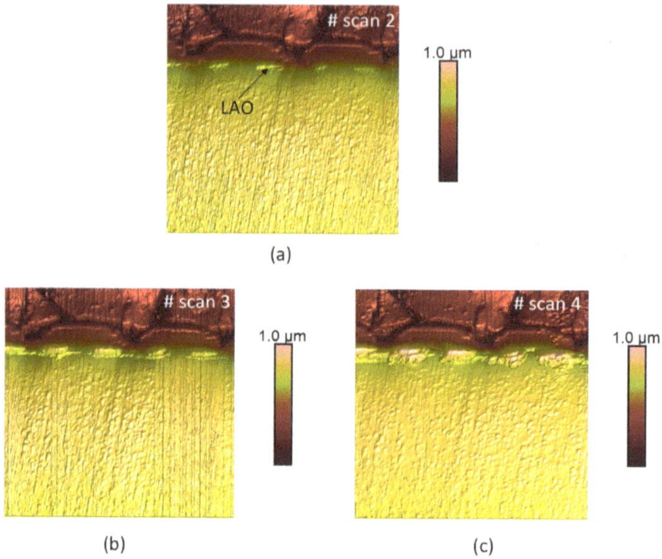

Figure 11. Consecutive AFM topographies of a commercial SiC-MOSFET: (**a**) second, (**b**) third and (**c**) fourth scan with $V_{DC} = -10$ V applied on the tip. Scan size: 14 µm × 14 µm.

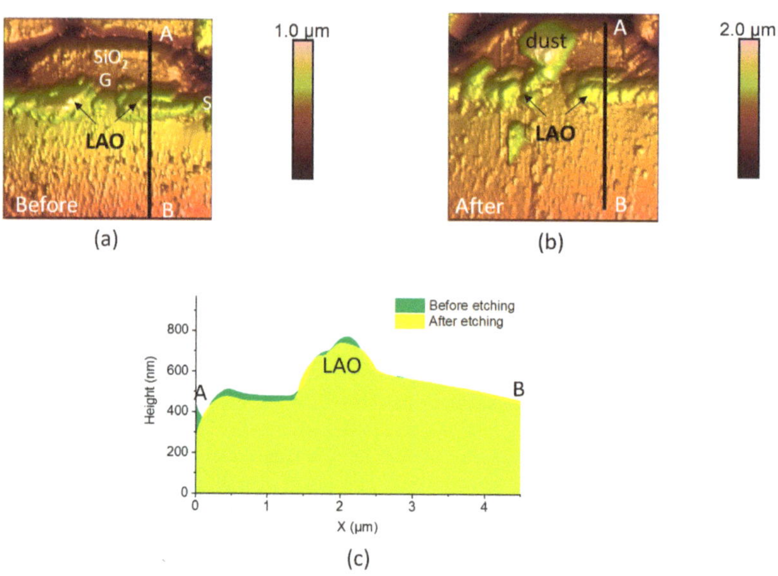

Figure 12. Topographies (**a**) before, (**b**) after the etching and (**c**) comparison of the profiles along the AB line after and before etching. The scan size is 6 µm × 6 µm. S is the source and G the gate of the SiC MOSFET.

5. Conclusions

The signature of the presence of a water meniscus on the AFM observation of SiC local electrical properties and the surface anodization was determined by FEM simulations and C-AFM measurements. The evaluation of the electric field and the current density highlights the influence of the water meniscus, but also of the oxide surface layer, as well as the doping level of the SiC. Being strongly dependent on the details of the water meniscus, the influence of extrinsic parameters, for example, the air humidity, is shown, explaining the scattering of reported experimental results. The simulation results are confirmed by AFM measurements, where the influence of the doping level and its threshold value of 7×10^{18} at/cm^3 for anodization was identified on epitaxial SiC staircase samples with gradual doping. For the COTS SiC MOSFETs, the anodization is present only on the highly doped areas of this complex structure, which can therefore be identified by simple topographical measurements after the application of a sufficiently high field through the AFM tip. The auto-limiting effect LAO oxide growth is also characterized by the study of the consecutive scans. However, the SiOCH nature of the oxide grown by surface anodization was demonstrated by selective KOH etching. This investigation highlights the LAO effect for doped SiC and demonstrates that this effect must be taken into account during electrical AFM measurements, in particular, in the case of successive scans This result could also provide a basis for the design and fabrication of AFM-oxidized nanostructures, using doping as an oxidation mask.

Author Contributions: Conceptualization, V.S., P.M.G. and R.C.G.; Methodology, C.V.-F., V.S., P.M.G., U.L. and R.C.G.; Software, C.V.-F.; Validation, C.V.-F., U.L. and R.C.G.; Investigation, C.V.-F., A.B. and R.C.G.; Resources, V.S. and P.M.G.; Writing—original draft, C.V.-F., A.B., V.S., P.M.G., U.L. and R.C.G.; Writing—review & editing, C.V.-F., V.S., P.M.G., U.L. and R.C.G.; Supervision, R.C.G. All authors have read and agreed to the published version of the manuscript.

Funding: This research received no external funding.

Institutional Review Board Statement: Not applicable.

Informed Consent Statement: Not applicable.

Data Availability Statement: Not applicable.

Acknowledgments: The authors would like to thank UNICAEN for support in the framework of the international partner Laboratory (IPL) OPTI-SiC2050, EPSRC grant EP/P017363/1 for epitaxial growth and Cendrine Folton for the confocal views.

Conflicts of Interest: The authors declare no conflict of interest.

References

1. Kumar, A.; Moradpour, M.; Losito, M.; Franke, W.-T.; Ramasamy, S.; Baccoli, R.; Gatto, G. Wide Band Gap Devices and Their Application in Power Electronics. *Energies* **2022**, *15*, 9172. [CrossRef]
2. Neudeck, P.G.; Spry, D.J.; Chen, L.; Prokop, N.F.; Krasowski, M.J. Demonstration of 4H-SiC Digital Integrated Circuits Above 800 °C. *IEEE Electron Device Lett.* **2017**, *38*, 1082–1085. [CrossRef]
3. Casady, J.; Johnson, R. Status of silicon carbide (SiC) as a wide-bandgap semiconductor for high-temperature applications: A review. *Solid-State Electron.* **1996**, *39*, 1409–1422. [CrossRef]
4. Umegami, H.; Harada, T.; Nakahara, K. Performance Comparison of Si IGBT and SiC MOSFET Power Module Driving IPMSM or IM under WLTC. *World Electr. Veh. J.* **2023**, *14*, 112. [CrossRef]
5. High-Voltage SiC Power Devices for Improved Energy Efficiency. Available online: https://www.jstage.jst.go.jp/article/pjab/98/4/98_PJA9804B-02/_article (accessed on 2 June 2023).
6. Murray, H.; Germanicus, R.; Doukkali, A.; Martin, P.; Domenges, B.; Descamps, P. Analytic description of scanning capacitance microscopy. *J. Vac. Sci. Technol. B Microelectron. Nanometer Struct. Process. Meas. Phenom.* **2007**, *25*, 1340–1352. [CrossRef]
7. Suda, J.; Nakamura, S.-I.; Miura, M.; Kimoto, T.; Matsunami, H. Scanning Capacitance and Spreading Resistance Microscopy of SiC Multiple-pn-Junction Structure. *Jpn. J. Appl. Phys.* **2002**, *41*, L40. [CrossRef]
8. Cho, Y.C.Y.; Kazuta, S.K.S.; Ohara, K.O.K.; Odagawa, H.O.H. Quantitative Measurement of Linear and Nonlinear Dielectric Characteristics Using Scanning Nonlinear Dielectric Microscopy. *Jpn. J. Appl. Phys.* **2000**, *39*, 3086. [CrossRef]
9. Chinone, N.; Cho, Y. Visualization of Gate-Bias-Induced Carrier Redistribution in SiC Power DIMOSFET Using Scanning Nonlinear Dielectric Microscopy. *IEEE Trans. Electron Devices* **2016**, *63*, 3165–3170. [CrossRef]

10. Yamasue, K.; Cho, Y. Surface Potential Fluctuations of SiO_2/SiC Interfaces Investigated by Local Capacitance-Voltage Profiling Based on Time-Resolved Scanning Nonlinear Dielectric Microscopy. *Mater. Sci. Forum* **2022**, *1062*, 335–340. [CrossRef]
11. Giannazzo, F.; Shtepliuk, I.; Ivanov, I.G.; Iakimov, T.; Kakanakova-Georgieva, A.; Schilirò, E.; Fiorenza, P.; Yakimova, R. Probing the uniformity of hydrogen intercalation in quasi-free-standing epitaxial graphene on SiC by micro-Raman mapping and conductive atomic force microscopy. *Nanotechnology* **2019**, *30*, 284003. [CrossRef]
12. Eriksson, J.; Roccaforte, F.; Giannazzo, F.; Nigro, R.L.; Raineri, V.; Lorenzzi, J.; Ferro, G. Improved Ni/3C-SiC contacts by effective contact area and conductivity increases at the nanoscale. *Appl. Phys. Lett.* **2009**, *94*, 112104. [CrossRef]
13. Giannazzo, F.; Fiorenza, P.; Saggio, M.; Roccaforte, F. Nanoscale Characterization of SiC Interfaces and Devices. *Mater. Sci. Forum* **2014**, *778–780*, 407–413. [CrossRef]
14. Fiorenza, P.; Schilirò, E.; Giannazzo, F.; Bongiorno, C.; Zielinski, M.; La Via, F.; Roccaforte, F. On the origin of the premature breakdown of thermal oxide on 3C-SiC probed by electrical scanning probe microscopy. *Appl. Surf. Sci.* **2020**, *526*, 146656. [CrossRef]
15. Park, G.; Choi, Y.; Shin, S.; Lee, Y.; Hong, S. Nanoscale Visualization of the Electron Conduction Channel in the SiO/Graphite Composite Anode. *ACS Appl. Mater. Interfaces* **2022**, *14*, 30639–30648. [CrossRef]
16. Elpelt, R.; Zippelius, B.; Doering, S.; Winkler, U. Employing Scanning Spreading Resistance Microscopy (SSRM) for Improving TCAD Simulation Accuracy of Silicon Carbide. *Mater. Sci. Forum* **2017**, *897*, 295–298. [CrossRef]
17. Germanicus, R.C.; Leclère, P.; Guhel, Y.; Boudart, B.; Touboul, A.D.; Descamps, P.; Hug, E.; Eyben, P. On the effects of a pressure induced amorphous silicon layer on consecutive spreading resistance microscopy scans of doped silicon. *J. Appl. Phys.* **2015**, *117*, 244306. [CrossRef]
18. Song, X.; Michaud, J.F.; Cayrel, F.; Zielinski, M.; Portail, M.; Chassagne, T.; Collard, E.; Alquier, D. Evidence of electrical activity of extended defects in 3C–SiC grown on Si. *Appl. Phys. Lett.* **2010**, *96*, 142104. [CrossRef]
19. Chang, A.S.; Li, B.; Wang, S.; Frisone, S.; Goldman, R.S.; Han, J.; Lauhon, L.J. Unveiling the influence of selective-area-regrowth interfaces on local electronic properties of GaN p-n junctions for efficient power devices. *Nano Energy* **2022**, *102*, 107689. [CrossRef]
20. Germanicus, R.C.; De Wolf, P.; Lallemand, F.; Bunel, C.; Bardy, S.; Murray, H.; Lüders, U. Mapping of integrated PIN diodes with a 3D architecture by scanning microwave impedance microscopy and dynamic spectroscopy. *Beilstein J. Nanotechnol.* **2020**, *11*, 1764–1775. [CrossRef]
21. Millán-Barba, J.; Bakkali, H.; Lloret, F.; Gutiérrez, M.; de Villoria, R.G.; Domínguez, M.; Haenen, K.; Araujo, D. Boron-doped diamond growth on carbon fibre: Enhancing the electrical conductivity. *Appl. Surf. Sci.* **2023**, *615*, 156382. [CrossRef]
22. Rius, G.; Lorenzoni, M.; Matsui, S.; Tanemura, M.; Perez-Murano, F. Boosting the local anodic oxidation of silicon through carbon nanofiber atomic force microscopy probes. *Beilstein J. Nanotechnol.* **2015**, *6*, 215–222. [CrossRef] [PubMed]
23. Gordon, A.E.; Fayfield, R.T.; Litfin, D.D.; Higman, T.K. Mechanisms of surface anodization produced by scanning probe microscopes. *J. Vac. Sci. Technol. B Microelectron. Nanometer Struct. Process. Meas. Phenom.* **1995**, *13*, 2805–2808. [CrossRef]
24. Ma, Y.-R.; Yu, C.; Yao, Y.-D.; Liou, Y.; Lee, S.-F. Tip-induced local anodic oxidation on the native SiO_2 layer of Si(111) using an atomic force microscope. *Phys. Rev. B* **2001**, *64*, 195324. [CrossRef]
25. Jo, Y.-D.; Seo, S.-H.; Bahng, W.; Kim, S.-C.; Kim, N.-K.; Koo, S.-M. Improved local oxidation of silicon carbide using atomic force microscopy. *Appl. Phys. Lett.* **2010**, *96*, 082105. [CrossRef]
26. Gao, J.; Luo, X.; Chang, W.; Qin, Y.; Xie, W.; Fan, P.; Hasan, R.M.M.; Geng, Y.; Yan, Y. Three-Dimensional Nanostructures Enabled by Customised Voltage Waveform-Induced Local Anodic Oxidation Lithography. In Proceedings of the 22nd EUSPEN International Conference and Exhibition, Geneva, Switzerland, 30 May–3 June 2022; pp. 285–288.
27. Ryu, Y.K.; Knoll, A.W. Oxidation and Thermal Scanning Probe Lithography for High-Resolution Nanopatterning and Nanodevices. Electrical Atomic Force Microscopy for Nanoelectronics. In *NanoScience and Technology*; Celano, U., Ed.; Springer International Publishing: Cham, Switzerland, 2019; pp. 143–172. [CrossRef]
28. Colangelo, F.; Piazza, V.; Coletti, C.; Roddaro, S.; Beltram, F.; Pingue, P. Local anodic oxidation on hydrogen-intercalated graphene layers: Oxide composition analysis and role of the silicon carbide substrate. *Nanotechnology* **2017**, *28*, 105709. [CrossRef]
29. Xie, X.N.; Chung, H.J.; Sow, C.H.; Wee, A.T.S. Native oxide decomposition and local oxidation of 6H-SiC (0001) surface by atomic force microscopy. *Appl. Phys. Lett.* **2004**, *84*, 4914–4916. [CrossRef]
30. Lorenzoni, M.; Torre, B. Scanning probe oxidation of SiC, fabrication possibilities and kinetics considerations. *Appl. Phys. Lett.* **2013**, *103*, 163109. [CrossRef]
31. Lee, J.H.; Ahn, J.J.; Hallén, A.; Zetterling, C.M.; Koo, S.M. Local Anodic Oxidation of Phosporous-Implanted 4H-SiC by Atomic Force Microscopy. *Mater. Sci. Forum* **2012**, *717–720*, 905–908. [CrossRef]
32. Germanicus, R.C.; Jouha, W.; Moultif, N.; De Wolf, P.; Shah, V.A.; Gammon, P.M.; Luders, U.; Latry, O. Parametric nano-electrical analysis for SiC junctions of a packaged device. In Proceedings of the 2022 IEEE Workshop on Wide Bandgap Power Devices and Applications in Europe (WiPDA Europe), Coventry, UK, 18–20 September 2022; pp. 1–6. [CrossRef]
33. Ryu, Y.K.; Garcia, R. Advanced oxidation scanning probe lithography. *Nanotechnology* **2017**, *28*, 142003. [CrossRef]
34. Liu, H.; Hoeppener, S.; Schubert, U.S. Nanoscale Materials Patterning by Local Electrochemical Lithography. *Adv. Eng. Mater.* **2016**, *18*, 890–902. [CrossRef]
35. Villeneuve-Faure, C.; Makasheva, K.; Boudou, L.; Teyssedre, G. Charge injection in thin dielectric layers by atomic force microscopy: Influence of geometry and material work function of the AFM tip on the injection process. *Nanotechnology* **2016**, *27*, 245702. [CrossRef]

36. Xi, J.; Liu, C.; Morgan, D.; Szlufarska, I. An Unexpected Role of H During SiC Corrosion in Water. *J. Phys. Chem. C* **2020**, *124*, 9394–9400. [CrossRef]
37. HButt, H.-J.; Cappella, B.; Kappl, M. Force measurements with the atomic force microscope: Technique, interpretation and applications. *Surf. Sci. Rep.* **2005**, *59*, 1–152. [CrossRef]
38. Cappella, B.; Dietler, G. Force-distance curves by atomic force microscopy. *Surf. Sci. Rep.* **1999**, *34*, 1–104. [CrossRef]
39. Rozhok, S.; Sun, P.; Piner, R.; Lieberman, M.; Mirkin, C.A. AFM Study of Water Meniscus Formation between an AFM Tip and NaCl Substrate. *J. Phys. Chem. B* **2004**, *108*, 7814–7819. [CrossRef]
40. Bartošík, M.; Kormoš, L.; Flajšman, L.; Kalousek, R.; Mach, J.; Lišková, Z.; Nezval, D.; Švarc, V.; Šamořil, T.; Šikola, T. Nanometer-Sized Water Bridge and Pull-Off Force in AFM at Different Relative Humidities: Reproducibility Measurement and Model Based on Surface Tension Change. *J. Phys. Chem. B* **2017**, *121*, 610–619. [CrossRef]
41. Sirghi, L.; Szoszkiewicz, R.; Riedo, E. Volume of a Nanoscale Water Bridge. *Langmuir* **2006**, *22*, 1093–1098. [CrossRef]
42. Cabrera, N.; Mott, N.F. Theory of the oxidation of metals. *Rep. Prog. Phys.* **1949**, *12*, 163. [CrossRef]
43. NNasir, N.M.; Shah, C.; Leech, P.; Reeves, G.; Pirogova, E.; Istivan, T.; Tanner, P.; Holland, A. Fabrication of 3C-silicon carbide membranes: Towards development of novel microdevices for biomedical applications. In Proceedings of the 2012 International Conference on Biomedical Engineering (ICoBE), Penang, Malaysia, 27–28 February 2012; pp. 589–593. [CrossRef]

materials

MDPI

Article

Analysis of Local Properties and Performance of Bilayer Epitaxial Graphene Field Effect Transistors on SiC

Dalal Fadil [1,2,*], Wlodek Strupinski [3], Emiliano Pallecchi [1] and Henri Happy [1,*]

1 University of Lille—IEMN CNRS UMR 8520, Avenue Poincaré, CS 60069, 59652 Villeneuve d'Ascq, France; emiliano.pallecchi@univ-lille.fr
2 Departament d'Enginyeria Electrònica, Universitat Rovira I Virgili, 43007 Tarragona, Spain
3 Faculty of Physics, Warsaw University of Technology, Koszykowa 75 Str., 00-662 Warsaw, Poland; wlodek.strupinski@pw.edu.pl
* Correspondence: dalal.fadil@urv.cat (D.F.); henri.happy@univ-lille.fr (H.H.)

Abstract: Epitaxial bilayer graphene, grown by chemical vapor deposition on SiC substrates without silicon sublimation, is crucial material for graphene field effect transistors (GFETs). Rigorous characterization methods, such as atomic force microscopy and Raman spectroscopy, confirm the exceptional quality of this graphene. Post-nanofabrication, extensive evaluation of DC and high-frequency properties enable the extraction of critical parameters such as the current gain (f_{max}) and cut-off frequency (f_t) of hundred transistors. The Raman spectra analysis provides insights into material property, which correlate with Hall mobilities, carrier densities, contact resistance and sheet resistance and highlights graphene's intrinsic properties. The GFETs' performance displays dispersion, as confirmed through the characterization of multiple transistors. Since the Raman analysis shows relatively homogeneous surface, the variation in Hall mobility, carrier densities and contact resistance cross the wafer suggest that the dispersion of GFET transistor's performance could be related to the process of fabrication. Such insights are especially critical in integrated circuits, where consistent transistor performance is vital due to the presence of circuit elements like inductance, capacitance and coplanar waveguides often distributed across the same wafer.

Keywords: bilayer graphene; silicon carbide; field effect transistors; nanofabrication; Raman spectra analysis; DC and RF characterizations

Citation: Fadil, D.; Strupinski, W.; Pallecchi, E.; Happy, H. Analysis of Local Properties and Performance of Bilayer Epitaxial Graphene Field Effect Transistors on SiC. *Materials* **2024**, *17*, 3553. https://doi.org/10.3390/ma17143553

Academic Editors: Marilena Vivona and Mike Jennings

Received: 5 June 2024
Revised: 8 July 2024
Accepted: 11 July 2024
Published: 18 July 2024

1. Introduction

Graphene is considered one of the most famous two-dimensional (2D) materials of this century [1–4]. This honeycomb carbon atom lattice exhibits extraordinary electrical [5,6], optical [7], thermal [8] and mechanical [9] properties and attracts huge attention for a wide panel of device applications, especially for radio-frequency transistors on rigid or on flexible substrate [4,10–12]. Since 2004, researchers continue improving the technique of growth, either by exfoliation [13–15] or by chemical vapor deposition (CVD). The most common technique used to grow graphene is using CVD on copper and transferring the carbon monolayer to a host substrate [16,17]. However, efforts remain to be made regarding the reliability and the reproducibility of the quality after transfer due to the cracks, wrinkles and residues, as reported by Smith et al. [18]. Another alternative is to grow graphene by CVD directly on SiC substrate by graphitization or by epitaxial growth without SiC sublimation or by hydrogen intercalation [19–25]. Besides the large-scale fabrication, the main advantage of CVD on SiC is to avoid the transfer from copper and improve the electronic properties of the material by using hydrogen intercalation, as reported by Ciuk et al. [26,27]. One of the advantages of the bilayer graphene on SiC substrate is overcoming the lack of a band gap; other possibilities reported in the literature include introducing defects, doping, strain and chemical bounding to the substrate [28–30]. Furthermore, the bilayer graphene field effect transistors (GFETs) on SiC have demonstrated better performances than monolayer

GFETs [31]. The 100 nm gate length (L_g) and 2 × 4 μm channel length (W_g) transistor exhibits 60 GHz of extrinsic current gain cut-off frequency f_{T_extr} and 25 GHz of maximum frequency of oscillation f_{max_extr}, and by annealing the bilayer graphene in hydrogen, the performance can be enhanced and reach f_{T_extr} = 70 GHz and f_{max_extr} = 120 GHz for L_g = 60 nm and W_g = 2 × 8 μm [32]. To analyze graphene properties, the standard non-destructive techniques used are (i) the atomic force microscopy AFM that give information about surface morphology including defects like holes, wrinkles or grain boundaries (ii) Raman spectroscopy which provides information about the number of layers, doping, strain and disorders in the graphene surface [33,34]. A new specific technique is non-contact Terahertz time domain spectroscopy. This complementary technique gives large scale information about mobility and carrier density [35].

The motivation behind this study goes beyond evaluating the RF performance of a single transistor. It also emphasizes the reproducibility of transistor performance across a large area, investigating whether the growth was uniform and the process reproducible. Understanding the material's properties and analyzing the local properties, mobility, contact resistance and their impact on device performance is crucial for enhancing the reliability and uniformity of transistors in integrated circuits.

In this work, we report the fabrication process of bilayer graphene field effect transistors. Initially, we verify the quality of the graphene samples before fabrication using AFM and Raman spectroscopy. Post-fabrication, we conduct further analysis using Raman spectroscopy, Hall mobility and TLM measurements to assess the local properties and the homogeneity of the samples. To evaluate device performance across the entire wafer, we performed DC and RF characterizations on hundreds of transistors. The RF performance data were then compared with the local properties of the samples, supporting a conclusion about device variability across the wafer. Our findings demonstrate significant performance variation across the wafer. The average $<f_t>/<f_{max}>$ ranged from 3/0.5 GHz in cell 4 to 16/11 GHz in cell 7 for devices with L_g = 200 nm and W_g = 2 × 30 μm. Notably, the devices in cell 7 and 8, which exhibited the best performance, also had the lowest contact resistance. This work highlights the important role of the fabrication process in the RF performance variability of graphene FETs fabricated from high-quality epitaxial bilayer graphene on a SiC substrate.

2. Materials and Methods

Epitaxial bilayer graphene was synthesized by chemical vapor deposition on a 500 μm thick, high-resistivity 6H-SiC (0001) substrate using a commercial horizontal CVD hot wall Aixtron VP508 reactor (Aixtron, Herzogenrath, Germany) equipped with an RF generator for heating. Prior to growth, in situ etching of the SiC surface was performed under a hydrogen atmosphere at 1600 °C and a chamber pressure of 100 mbar. The carbon films were deposited using propane as the carbon precursor. Our method employs high-temperature and low-argon-pressure CVD, creating laminar argon flow dynamics to protect the SiC substrate from silicon sublimation and facilitate propane mass transport, thereby enabling graphene epitaxy, as detailed in references [22,25,26]. The growth process was followed by in situ hydrogen intercalation at 1000 °C in a 900 mbar Ar atmosphere. The optimization of growth parameters aimed at achieving uniform bilayer graphene. Before fabrication, initial carrier density and the mobility was provided by Ciuk et al. to be around +8.3 × 10^{12} cm^{-2} and 850 $cm^2{\cdot}V^{-1}s^{-1}$, respectively. In contrast to the methodology described by P. Wehrfritz et al. [36], which utilized a similar SiC substrate, our process involves unique conditions including high-temperature in situ hydrogen etching and controlled low-argon pressure to prevent silicon sublimation. These optimizations contribute to the superior uniformity and controlled thickness of our bilayer graphene compared to prior studies. Notably, our approach introduces an innovative in situ hydrogen intercalation step at 1000 C, enhancing graphene quality, which was not emphasized in previous literature [36].

The AFM Bruken Icon model was used on mode tapping to determine the properties, the material and image of the surface. Image analysis was performed with WSxM5.0 Develop 8.3, a free software. High-resolution electron beam lithography VISTEC EBPG5000Plus was used for device fabrication. To analyze the local properties of the material, we used a HORIBA Jobin–Yvon lab system for Raman spectroscopy at a laser wavelength of 473 nm, using a 1 μm laser spot size and filters to deliver power less than 0.1 mW and ×100 objective lens to measure the different positions of the sample. HL5500PC was used to carry out Hall measurement. To measure the DC and RF performance of hundreds of devices, we used a standard probe station with Microtech's probes, Semiconductor Analyzer HP4155A, the Vector Network Analyzer HP4155A and the vector Network Analyzer Rohde & Schwarz ZVA67.

A. Before fabrication

Primary characterization was performed before fabrication. Figure 1a represents atomic force microscopy (AFM) images of a 60 × 60 μm^2 surface area of graphene observed at room temperature. The AFM images reveal well-oriented, parallel atomic steps with SiC terraces approximately 10 μm wide, separated by steps estimated to be a few nanometers in height. This phenomenon, known as step bunching, occurs during the growth-preceding in situ hydrogen etching of the SiC surface, as described in reference [27]. Within these terraces, Figure 1b showcases two distinct surface morphologies. The graphene roughness within individual terraces has a root mean square (RMS) value of 0.273 nm (Figure 1b, left) and 0.2406 nm (Figure 1b, right), significantly smoother than the 1 nm roughness reported in previous studies [36]. Figure 1c shows the SEM image of epitaxial graphene on the SiC substrate. We observed the graphene surface nucleation on the SiC steps. An example of a Raman spectroscopy of graphene on SiC and after extracting SiC peaks is presented in Figure 1d. We observed small intensity of the D peak compared to the G and 2D peaks, indicating a small disorder and defects in the material. Previous work about Raman studies in graphene on SiO_2/Si show that the shape of the 2D peak is an indicator of the number of layers [37]. For Raman analysis, the nature of the growth (exfoliation, CVD graphitization, CVD without graphitization) and the type of the substrates (SiO_2 or SiC) can affect the peaks' position and the value of the full width at half maximum (FWHM), which enhance the need to have more results about bilayer graphene on SiC [26,38,39]. Here, the full width at the half maximum of the 2D peak is around 59 cm^{-1}, comparable to the 41–62 cm^{-1} reported in bilayer graphene on SiC and other substrates [25,37]. In bilayer graphene, the 2D peak is typically broader and upshifted compared to monolayer graphene. This broadening is due to the presence of an additional phonon mode in bilayer graphene, resulting in a more complex 2D band shape [37]. Before fabrication, Raman spectra was established in three different locations randomly selected on the SiC wafer and presented in Figure 1e. The Raman spectrum does not change with changing locations.

B. Device fabrication.

The field effect transistors based on graphene were fabricated on a 15 × 15 mm^2 SiC wafer. Figure 2a represents the layout of the device. Each level of the layout is represented by a different color or contrast and represents a fabrication step of the process. First, the process is fixed by defining the alignment marks. It follows the etching of the graphene channel and contacts as a hole for improving contact resistance, as previously reported in [41,42]. The source and drain contacts were obtained by the standard lift-off process after evaporating 1.5 nm of nickel and 30 nm of gold metals. Here, a thin layer of nickel (1.5 nm) was deposited before in order to improve the metal adhesion on the surface. The dual T-gate with gate length (L_g) were defined by using three layers of poly-meta-methacrylate (PMMA), as shown in Figure 2b, where three different thicknesses were defined: 160 nm thick at the bottom part of the T-gate, 720 nm thick at the top part of the T-gate and 130 nm at the resist followed by electron beam (e-beam) lithography. After the development of these multilayers' resists, the gate oxide is deposited using 2 nm of evaporated aluminum four times, followed by oxidation in ambient air for 24 h. Finally, the coplanar access

Ni (50 nm)/Au (300 nm) are deposited, followed by lift-off. The cross-section schema of the final transistor is presented in Figure 2c. The top view of the active part of the final dual-T-gate-transistor is illustrated in the scanning electron microscopy (SEM) image in Figure 2d (left) while the cross-section of the gate part was illustrated by the focused ion beam (FIB) technique in Figure 2d (right) and shows the shape of the final T-gate of the GFET. A picture of the 15 × 15 mm^2 final wafer where there are 8 cells and 458 transistors is shown in Figure 2e.

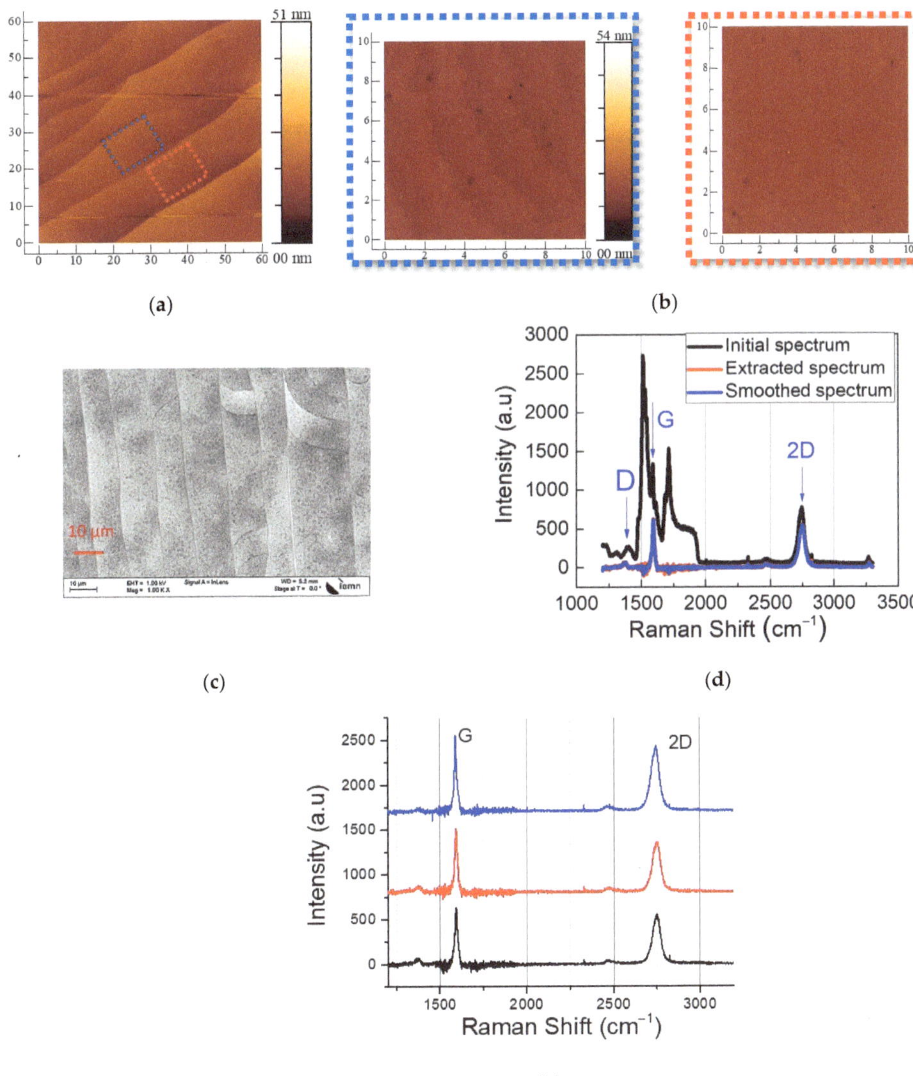

Figure 1. AFM images of bilayer graphene surface on 6H-SiC substrate [40]. (**a**) 60 × 60 µm^2 image and the blue and red dash square 10 × 10 µm^2 is represented in (**b**). (**c**) SEM image with 10 µm scale bare. (**d**) Raman spectroscopy of graphene on SiC substrate. The black trace is the spectrum of graphene and SiC, the red trace is the graphene spectrum once SiC Raman peaks are subtracted, the blue trace is the smoothed spectrum. (**e**) Raman spectra at different locations (blue, red, and black) on the SiC wafer.

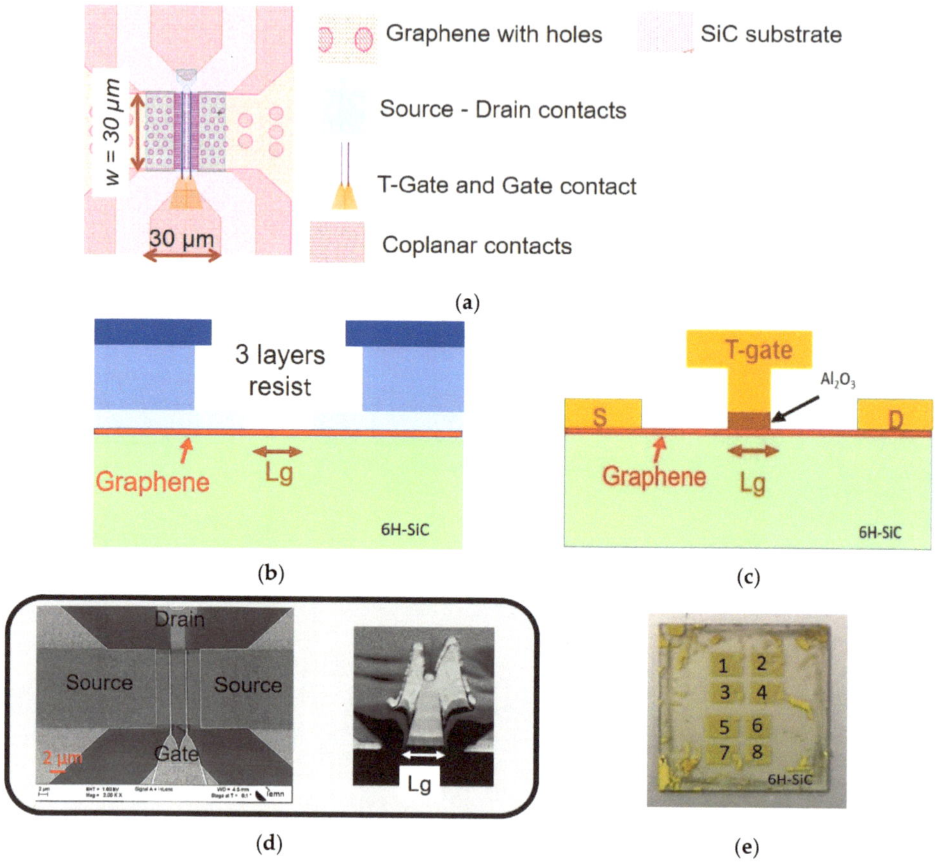

Figure 2. (**a**) A picture of the layout of the device. "+" is source-drain contact layer. (**b**) A schematic image representing the step of the realization of the T gate process with the three layers' resists. (**c**) A schematic of the side view of the T gate transistor. (**d**) (**d, left**) An SEM image of the transistor in the end of the process. (**d, right**) An FIB image of the transistor showing the T-gate structure, L_g = 200 nm. (**e**) A photograph of the final devices fabricated on the 15 mm × 15 mm SiC substrate, including 458 transistors and height cells numbered from 1 to 8.

3. Results and Discussion

After fabrication, we conducted an in-depth characterization of the electrical properties of graphene. The wafer was tested within patterned structures, including Hall measurements and transmission line measurement (TLM), as illustrated in Figure 3a, across various cells. Contact and sheet resistance, which represent the resistance as a function of the distance between contact points (0.5, 1, 2, 4, 16, 24 μm), were extracted from a linear fit using the TLM method. The *y*-axis intercept provides the contact resistance (2Rc), while the slope of the linear fit represents the Rsh ratio to the width of the graphene channel [41,43]. All extracted values of contact resistance and sheet resistance for each cell are presented in Figure 3d.

To complete the analysis, we acquired seven Raman spectra from each cell of the wafer and extracted an average Raman spectrum. Figure 3c summarizes the average Raman spectra representing each cell. Mobility values, carrier densities and Raman peaks are provided in the table in Figure 3d. It shows a dispersion of values for contact resistance, sheet resistance, Hall mobility and carrier density across the wafer. The highest Hall mobility (626–832 $cm^2/V{\cdot}s$), carrier densities 18–16 ($\times 10^{12}$ cm^{-2}), lowest values of contact resistance (678–650 $\Omega{\cdot}m$) and sheet resistance (323–286 Ω/sq) were obtained in the cells 7–8. However, the low value of mobility observed in cells 1–6 is possibly due to the high contact resistance. The ohmic contact between graphene and metal contacts exhibits large contact resistance, significantly reducing the apparent mobility of contacted graphene and hindering its potential in high-frequency applications [42].

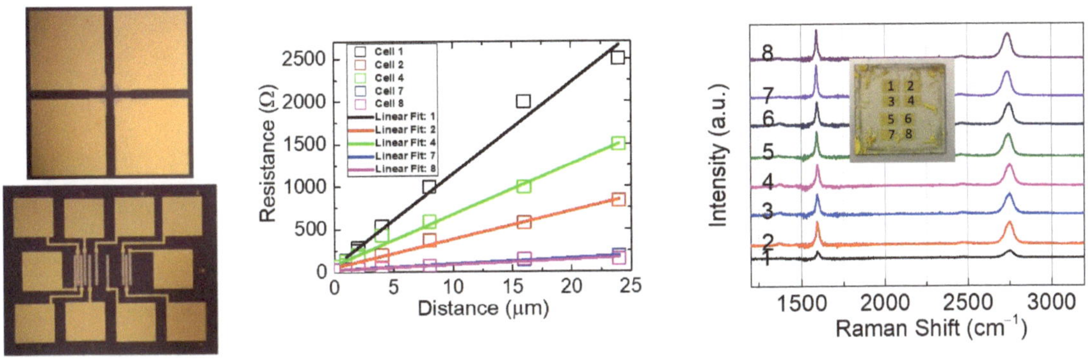

(a) (b) (c)

Cell No	Carrier densities x10¹² (cm⁻²)	Hall Mobility (cm²/V.s)	Contact resistance (Ω.μm)	Sheet Resistance (Ω/sq)	G peak position (cm⁻¹)	2D peak position (cm⁻¹)	FWHM (G)	FWHM (2D)
1	4	360	1689	5175	1591	2750	33	65
2	5	439	1335	1582	1590	2751	19	65
3	4	367	-	-	1590	2750	24	57
4	-	-	1971	2845	1593	2748	16	57
5	10	281	-	-	1590	2744	16	54
6	6	104	-	-	1592	2749	17	55
7	18	625	678	323	1590	2738	14	57
8	16	832	650	286	1591	2740	14	57

(d)

Figure 3. (**a**) An optical image of the Hall and TLM pattern. (**b**) The transmission line measurement (TLM) and the linear fit to extract the correspondent contact resistance and sheet resistance. (**c**) A recapitulation of the averaged Raman spectra of graphene on SiC measured in the eight cells of the wafer. (**d**) A table summarizing the values of the Hall mobility, carrier densities and Raman peaks' positions and the full width at the half maximum of the G and 2D peaks. Missing values are due to faulty components.

From the Raman analysis, the position of the G peak remains consistent across the wafer, while the full width at the half maximum of the G peak FWHM (G) varies. The thinnest FWHM (G) is observed in cells 7 and 8. The 2D peak exhibits a blueshift of approximately 20 cm^{-1}, from 2751 cm^{-1} in cell 2 to 2731 cm^{-1} in cell 7. This shift could be attributed to interlayer coupling effects in bilayer graphene. Ciuk et al. [25] investigated bilayer CVD graphene on SiC and observed the blueshift of the 2D peak and redshift of the

G peak due to the strain at the edges of the steps. Moreover, Das et al. [44] demonstrated that in graphene on SiO_2/Si, the position of the G peak increased, while the 2D peak and the FWHM (G) decreased with increased dopant concentration. Ferrari et al. [37] showed that the electron and hole doping upshifts and sharpens the G peak. Additionally, graphene can be doped chemically during fabrication, affecting carrier densities, 2D-peak width and position. Previous studies have shown that doping increases the 2D position due to the modification of lattice parameters, which modifies the total number of charges and leads to a stiffening/softening of the phonons [45]. The analysis of the table in Figure 3d clearly shows that cells 7–8 present the highest mobility, carrier densities, lowest contact and sheet resistance and smallest FWHM (G), suggesting favorable prospects for effective device and circuit applications.

To evaluate the performance of our devices, DC and RF measurements were performed using a standard probe station with Microtech's probes, Semiconductor Analyzer HP4155A, the Vector Network Analyzer HP4155A and the vector Network Analyzer Rohde & Schwarz ZVA67. A common calibration procedure of Line-Reflect-Reflect-Match (LRRM) for RF measurements was established before measurements. In total, 195 transistors of the GFETs were measured.

Figure 4 represents an example of the electrical characterization of a transistor in cell 7 (Figure 2e), with a 200 nm gate length and a 30 × 2 µm channel width. The transfer characteristic I_{DS} as function of V_{GS} at V_{DS} = 1.5 V and the transconductance ($g_m = dI_{DS}/dV_{GS}$) are shown in Figure 4a. The I_{DS}–V_{GS} curve is non-monotonic. As V_{GS} changes from −3 V to +3 V, I_{DS} shows a change in slopes initially decreasing, then reversing direction. This non-monotonic behavior is accurately reflected in the g_m curve, calculated as the derivative of I_{DS} as a function of V_{GS}. The characteristic of decreasing and then increasing g_m is consistent with our observations and not indicative of bipolar behavior as observed in [46]. The on/off extracted is approximately 1.4, within the V_{GS} range of ±3 V. The peak g_m reaches 4.6 mS (76 mS/mm) at V_{GS} = 1.3 V. The Dirac point is located at V_{GS} higher than 3 V, suggesting p-type doping in the graphene channel. Figure 4b shows the output characteristics, sweeping the gate voltage from −2 V to +2 V in 0.5 V steps. The maximum current is around 53 mA (0.88 mA/µm) at V_{DS} = +1.5 V and V_{GS} = −2 V, with no observed saturation current. This work investigates graphene's potential for high-frequency analog applications, focusing on its high-frequency characteristics rather than switching capability and looking for it to reach the saturation current. Graphene's semi-metallic nature results in a low on/off ratio, which is less critical for RF amplifiers compared to synaptic transistors or for digital applications [12,46]. Due to its high carrier mobility and conductivity, graphene excels in GHz frequency operations, maintaining steady current mobility flow at specific bias points [47].

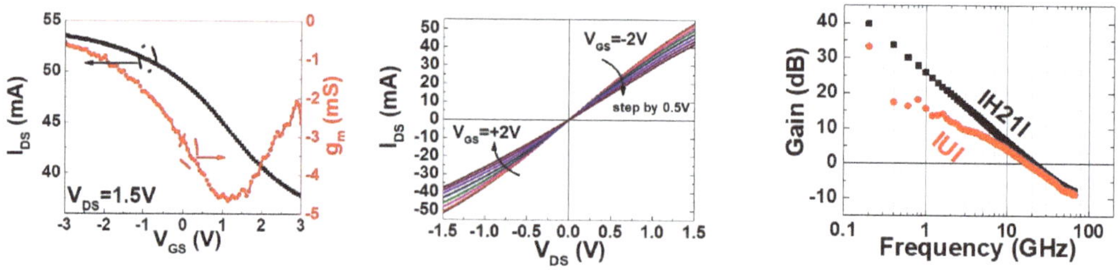

(a) (b) (c)

Figure 4. Example of DC and RF characteristics of dual T-gate graphene transistor in cell 7 having gate channel length (L_g) equal to 200 nm and dual-gate width (w) 2 × 30 µm. (**a**) DC measurement I_{DS}–V_{GS} and g_m–V_{GS}. (**b**) The voltage transfer characteristics as a function of V_{GS} varying from −2 V to +2 V by 0.5 V steps. (**c**) RF characteristic includes the as-measured values of the current gain H_{21} and the unilateral power gain U as a function of the frequency at V_{DS} = 1.5 V and V_{GS} = 1.3 V. The cut off frequency f_t and the maximum oscillation frequency f_{max} have been extracted [40].

The extrinsic RF characterization is reported in Figure 4c. It shows the current gain H_{21} and the Mason's gain U of the device in a frequency range of 0.6 to 67 GHz. The maximum frequency of oscillation (f_{max}) and the current gain cut-off frequency (f_t) are, respectively, the frequency at which the power gain (U) and the current gain (H_{21}) are equal to 1. The on-probe measurement of the cut-off frequency and the maximum oscillation frequency reveal, respectively, f_t = 21 GHz and f_{max} = 18 GHz at V_{GS} = +1 V and the gain curves follow the slope of 20 dB/decade, as expected. The extrinsic value of the performance is obtained after the capacitances related to the length of the transmission line of transistor access are removed [48]. They are, respectively, $f_{T\text{-}extr}$ = 62 GHz and $f_{max\text{-}extr}$ remains the same.

These values are comparable to the recent value achieved by [32] in the bilayer graphene on SiC.

To understand the evolution of RF performance across the eight cells of the full wafer shown in Figure 5a, we analyzed the average values of f_t, f_{max} and the optimum gate voltage for each individual cell. The optimum gate voltages for RF measurement were determined from the g_m—V_{GS} characteristic, where the transconductance is at its maximum. The average gate voltage $<V_{GS}>$ is shown in Figure 5b and represents the mean gate voltage computed in each cell. The average cut-off frequency $<f_t>$ and maximum oscillation frequency $<f_{max}>$ are reported in Figure 5c,d, respectively. They clearly show that the best performance is achieved in cells 7 and 8, with $<f_t>$ of 16–14 GHz and $<f_{max}>$ of 11–9 GHz, respectively. Figure 5e summarizes the f_t and f_{max} values for each cell as a function of the optimum gate voltage. Interestingly, the best transistor performances are obtained in cells 7 and 8 for gate voltages of at least 1 V. This variation in gate voltage may be attributed to the variation of the dopant across the wafer as mentioned in the Raman analysis. Additionally, differences in contact resistance across various cells contribute to the observed variability. To mitigate this issue, implementing a robust cleaning process is essential to eliminate polymer contamination during fabrication. This approach will reduce contact resistance, enhance uniformity, and minimize performance variability. It is important to acknowledge that both contact resistance and graphene quality are important for overall device performance. However, our study suggests that the RF performance is immediately limited by contact resistance related to the process of fabrication.

The findings highlight the critical role of the fabrication process in the RF performance variability of graphene FETs fabricated from high-quality epitaxial bilayer graphene on a SiC substrate. Despite the material's exceptional quality, confirmed through AFM and Raman spectroscopy, the observed dispersion in values of Hall mobility, carrier densities and contact resistance across the wafer suggest that these performance discrepancies result from the fabrication stages rather than inherent material inconsistencies. This understanding underscores the need for refined manufacturing techniques to ensure consistent transistor behavior, which is essential for the reliable operation of integrated circuits that rely on uniform performance of GFET in wafer scale. The insights gained from this study are essential for advancing GFET technology, particularly in high-frequency applications where reproducible device characteristics are crucial.

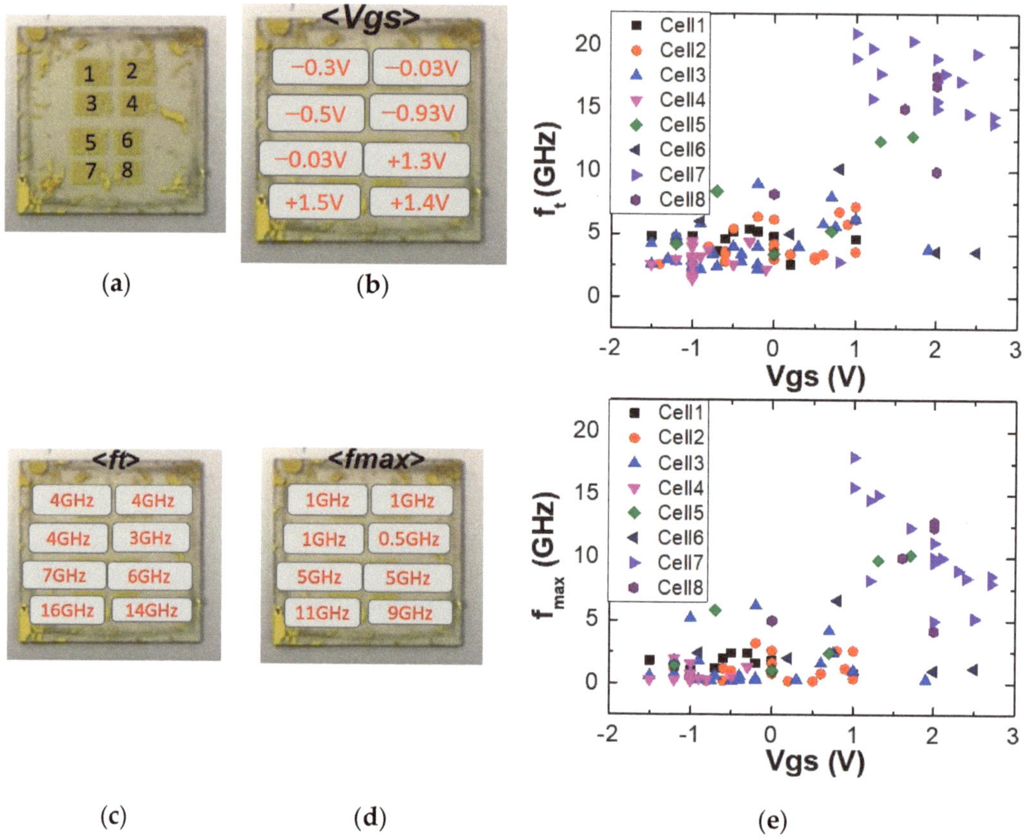

Figure 5. (**a**) Photography of the 15 mm × 15 mm graphene wafer presented previously in Figure 2e. (**b**) V_{GS} is the bias gate voltage related the maximum the transconductance and where the best performance of the transistor is expected. $<V_{GS}>$ is the average value of all the gate voltage measured in each cell. (**c**,**d**) Representation of the average values of the cut-off frequency $<f_t>$ and the maximum oscillation frequency $<f_{max}>$ computed for each cell. (**e**) Graph summarizing the evolution of the on-probe values of f_t and f_{max} as function of the biased gate voltage V_{GS} of the transistors in each cell.

4. Conclusions

In this paper, we fabricated hundreds of bilayer graphene field effect transistors on a SiC substrate. We provided Raman spectrum analysis combined with the Hall mobility, carrier densities, contact resistance and sheet resistance toward a 15 mm × 15 mm graphene wafer. The analyses of local Raman have shown that the 2D peak and the FWHM (G) changed, as well as the Hall mobility, carrier densities and contact resistance in the wafer. The analysis of the RF performances was performed and compared to the local properties of the wafer. It revealed dispersion of the performances and correlation between the RF performance, contact resistance and Hall mobilities. This work completes the few works about local Raman investigation on bilayer graphene on SiC and highlights the importance of local analyses of the properties of the material to evaluate the performance of electronic devices where large-scale, homogenous and high-quality fabrication of the material is needed for high-performance circuit.

Author Contributions: D.F. is the main author of this work. She assumed the design and the fabrication of the devices, the DC and RF characterizations, as well as the preparation of the graphs of the manuscript. W.S. provided the graphene material used in this work. The CVD bilayer graphene used was grown in his research group. E.P. and H.H. initiated this work and contributed to high-frequency characterizations and interpretations of the results. All authors reviewed and commented on the manuscript. All authors have read and agreed to the published version of the manuscript.

Funding: This research was funded by Horizon 2020 Framework Program: GrapheneCore2 785219 and GrapheneCore3 881603.

Institutional Review Board Statement: Not applicable.

Informed Consent Statement: Not applicable.

Data Availability Statement: Data is contained within the article.

Acknowledgments: The authors gratefully acknowledge financial support from the EU FP7-ICT-2013-FET-F GRAPHENE Flagship project (GrapheneCore2 no. 785219 and GrapheneCore3 n° 881603). This work was also partly supported by the French RENATECH network. D.F. is currently funded from the European Union's Horizon 2020 research and innovation program under the Marie Sklodowska Curie grant agreement (no. 101025770).

Conflicts of Interest: The authors declare no conflicts of interest.

References

1. Geim, A.K. Graphene: Status and Prospects. *Science* **2009**, *324*, 1530–1534. [CrossRef] [PubMed]
2. Kaul, A.B. Two-dimensional layered materials: Structure, properties, and prospects for device applications. *J. Mater. Res.* **2014**, *29*, 348–361. [CrossRef]
3. Ferrari, A.C.; Bonaccorso, F.; Fal'Ko, V.; Novoselov, K.S.; Roche, S.; Bøggild, P.; Borini, S.; Koppens, F.H.L.; Palermo, V.; Pugno, N.; et al. Science and technology roadmap for graphene, related two-dimensional crystals, and hybrid systems. *Nanoscale* **2015**, *7*, 4598–4810. [CrossRef] [PubMed]
4. Zhang, H.; Chhowalla, M.; Liu, Z. 2D nanomaterials: Graphene and transition metal dichalcogenides. *Chem. Soc. Rev.* **2018**, *47*, 3015–3017. [CrossRef] [PubMed]
5. Bolotin, K.; Sikes, K.; Jiang, Z.; Klima, M.; Fudenberg, G.; Hone, J.; Kim, P.; Stormer, H. Ultrahigh electron mobility in suspended graphene. *Solid State Commun.* **2008**, *146*, 351–355. [CrossRef]
6. Pizzocchero, F.; Gammelgaard, L.; Jessen, B.S.; Caridad, J.M.; Wang, L.; Hone, J.; Bøggild, P.; Booth, T.J. The hot pick-up technique for batch assembly of van der Waals heterostructures. *Nat. Commun.* **2016**, *7*, 11894. [CrossRef] [PubMed]
7. Koppens, F.H.L.; Mueller, T.; Avouris, P.; Ferrari, A.C.; Vitiello, M.S.; Polini, M. Photodetectors based on graphene, other two-dimensional materials and hybrid systems. *Nat. Nanotechnol.* **2014**, *9*, 780–793. [CrossRef] [PubMed]
8. Balandin, A.A. Thermal properties of graphene and nanostructured carbon materials. *Nat. Mater.* **2011**, *10*, 569–581. [CrossRef]
9. Lee, C.; Wei, X.; Kysar, J.W.; Hone, J. Measurement of the elastic properties and intrinsic strength of monolayer graphene. *Science* **2008**, *321*, 385–388. [CrossRef]
10. Schwierz, F. Graphene transistors. *Nat. Nanotechnol.* **2010**, *5*, 487–496. [CrossRef]
11. Wu, Y.; Jenkins, K.A.; Valdes-Garcia, A.; Farmer, D.B.; Zhu, Y.; Bol, A.A.; Dimitrakopoulos, C.; Zhu, W.; Xia, F.; Avouris, P.; et al. State-of-the-Art Graphene High-Frequency Electronics. *Nano Lett.* **2012**, *12*, 3062–3067. [CrossRef] [PubMed]
12. Schwierz, F. Graphene Transistors: Status, Prospects, and Problems. *Proc. IEEE* **2013**, *101*, 1567–1584. [CrossRef]
13. Novoselov, K.S.; Geim, A.K.; Morozov, S.V.; Jiang, D.; Zhang, Y.; Dubonos, S.V.; Grigorieva, I.V.; Firsov, A.A. Electric field effect in atomically thin carbon films. *Science* **2004**, *306*, 666–669. [CrossRef] [PubMed]
14. Novoselov, K.S.; Fal'Ko, V.I.; Colombo, L.; Gellert, P.R.; Schwab, M.G.; Kim, K. A roadmap for graphene. *Nature* **2012**, *490*, 192–200. [CrossRef] [PubMed]
15. Tielrooij, K.-J.; Hesp, N.C.H.; Principi, A.; Lundeberg, M.B.; Pogna, E.A.A.; Banszerus, L.; Mics, Z.; Massicotte, M.; Schmidt, P.; Davydovskaya, D.; et al. Out-of-plane heat transfer in van der Waals stacks through electron–hyperbolic phonon coupling. *Nat. Nanotechnol.* **2018**, *13*, 41–46. [CrossRef] [PubMed]
16. Kim, K.S.; Zhao, Y.; Jang, H.; Lee, S.Y.; Kim, J.M.; Kim, K.S.; Ahn, J.-H.; Kim, P.; Choi, J.-Y.; Hong, B.H. Large-scale pattern growth of graphene films for stretchable transparent electrodes. *Nature* **2009**, *457*, 706–710. [CrossRef] [PubMed]
17. Deokar, G.; Avila, J.; Razado-Colambo, I.; Codron, J.-L.; Boyaval, C.; Galopin, E.; Asensio, M.-C.; Vignaud, D. Towards high quality CVD graphene growth and transfer. *Carbon* **2015**, *89*, 82–92. [CrossRef]
18. Smith, A.D.; Wagner, S.; Kataria, S.; Malm, B.G.; Lemme, M.C.; Ostling, M. Wafer-Scale Statistical Analysis of Graphene FETs—Part I: Wafer-Scale Fabrication and Yield Analysis. *IEEE Trans. Electron. Devices* **2017**, *64*, 3919–3926. [CrossRef]
19. Berger, C.; Song, Z.M.; Li, T.B.; Li, X.B.; Ogbazghi, A.Y.; Feng, R.; Dai, Z.N.; Marchenkov, A.N.; Conrad, E.H.; First, P.N.; et al. Ultrathin Epitaxial Graphite: 2D Electron Gas Properties and a Route toward Graphene-based Nanoelectronics. *J. Phys. Chem. B* **2004**, *108*, 19912–19916. [CrossRef]

20. Wu, Y.Q.; Ye, P.D.; Capano, M.A.; Xuan, Y.; Sui, Y.; Qi, M.; Cooper, J.A.; Shen, T.; Pandey, D.; Prakash, G.; et al. Top-gated graphene field-effect-transistors formed by decomposition of SiC. *Appl. Phys. Lett.* **2008**, *92*, 092102. [CrossRef]
21. Ferrer, F.J.; Moreau, E.; Vignaud, D.; Godey, S.; Wallart, X. Atomic scale flattening, step formation and graphitization blocking on 6H- and 4H-SiC{0 0 0 1} surfaces under Si flux. *Semicond. Sci. Technol.* **2009**, *24*, 125014. [CrossRef]
22. Strupinski, W.; Grodecki, K.; Wysmolek, A.; Stepniewski, R.; Szkopek, T.; Gaskell, P.E.; Grüneis, A.; Haberer, D.; Bozek, R.; Krupka, J.; et al. Graphene Epitaxy by Chemical Vapor Deposition on SiC. *Nano Lett.* **2011**, *11*, 1786–1791. [CrossRef] [PubMed]
23. Guo, Z.; Dong, R.; Chakraborty, P.S.; Lourenco, N.; Palmer, J.; Hu, Y.; Ruan, M.; Hankinson, J.; Kunc, J.; Cressler, J.D.; et al. Record Maximum Oscillation Frequency in C-Face Epitaxial Graphene Transistors. *Nano Lett.* **2013**, *13*, 942–947. [CrossRef] [PubMed]
24. Pallecchi, E.; Lafont, F.; Cavaliere, V.; Schopfer, F.; Mailly, D.; Poirier, W.; Ouerghi, A. High Electron Mobility in Epitaxial Graphene on 4H-SiC(0001) via post-growth annealing under hydrogen. *Sci. Rep.* **2014**, *4*, 4558. [CrossRef] [PubMed]
25. Ciuk, T.; Cakmakyapan, S.; Ozbay, E.; Caban, P.; Grodecki, K.; Krajewska, A.; Pasternak, I.; Szmidt, J.; Strupinski, W. Step-edge-induced resistance anisotropy in quasi-free-standing bilayer chemical vapor deposition graphene on SiC. *J. Appl. Phys.* **2014**, *116*, 123708. [CrossRef]
26. Ciuk, T.; Caban, P.; Strupinski, W. Charge carrier concentration and offset voltage in quasi-free-standing monolayer chemical vapor deposition graphene on SiC. *Carbon* **2016**, *101*, 431–438. [CrossRef]
27. Ciuk, T.; Strupinski, W. Statistics of epitaxial graphene for Hall effect sensors. *Carbon* **2015**, *93*, 1042–1049. [CrossRef]
28. Zhou, S.; Gweon, G.-H.; Fedorov, A.; First, P.N.; De Heer, W.A.; Lee, D.-H.; Guinea, F.; Neto, A.H.C.; Lanzara, A. Substrate-induced bandgap opening in epitaxial graphene. *Nat. Mater.* **2007**, *6*, 770–775, Corrigendum in **2007**, *6*, 916. [CrossRef]
29. Zhang, Y.; Tang, T.-T.; Girit, C.; Hao, Z.; Martin, M.C.; Zettl, A.; Crommie, M.F.; Shen, Y.R.; Wang, F. Direct observation of a widely tunable bandgap in bilayer graphene. *Nature* **2009**, *459*, 820–823. [CrossRef]
30. Xu, X.; Liu, C.; Sun, Z.; Cao, T.; Zhang, Z.; Wang, E.; Liu, Z.; Liu, K. Interfacial engineering in graphene bandgap. *Chem. Soc. Rev.* **2018**, *47*, 3059–3099. [CrossRef]
31. He, Z.-Z.; Yang, K.-W.; Yu, C.; Liu, Q.-B.; Wang, J.-J.; Song, X.-B.; Han, T.-T.; Feng, Z.-H.; Cai, S.-J. Comparative Study of Monolayer and Bilayer Epitaxial Graphene Field-Effect Transistors on SiC Substrates. *Chin. Phys. Lett.* **2016**, *33*, 086801. [CrossRef]
32. Yu, C.; He, Z.Z.; Song, X.B.; Liu, Q.B.; Han, T.T.; Dun, S.B.; Wang, J.J.; Zhou, C.J.; Guo, J.C.; Lv, Y.J.; et al. Improvement of the Frequency Characteristics of Graphene Field-Effect Transistors on SiC Substrate. *IEEE Electron. Device Lett.* **2017**, *38*, 1339–1342. [CrossRef]
33. Neumann, C.; Reichardt, S.; Venezuela, P.; Drögeler, M.; Banszerus, L.; Schmitz, M.; Watanabe, K.; Taniguchi, T.; Mauri, F.; Beschoten, B.; et al. Raman spectroscopy as probe of nanometre-scale strain variations in graphene. *Nat. Commun.* **2015**, *6*, 8429. [CrossRef] [PubMed]
34. Long, F.; Yasaei, P.; Sanoj, R.; Yao, W.; Král, P.; Salehi-Khojin, A.; Shahbazian-Yassar, R. Characteristic Work Function Variations of Graphene Line Defects. *ACS Appl. Mater. Interfaces* **2016**, *8*, 18360–18366. [CrossRef] [PubMed]
35. Buron, J.D.; Pizzocchero, F.; Jepsen, P.U.; Petersen, D.H.; Caridad, J.M.; Jessen, B.S.; Booth, T.J.; Bøggild, P. Graphene mobility mapping. *Sci. Rep.* **2015**, *5*, 12305. [CrossRef] [PubMed]
36. Wehrfritz, P.; Fromm, F.; Malzer, S.; Seyller, T. Quasi-freestanding epitaxial graphene transistor with silicon nitride top gate. *J. Phys. D Appl. Phys.* **2014**, *47*, 305103. [CrossRef]
37. Ferrari, A.C. Raman spectroscopy of graphene and graphite: Disorder, electron–phonon coupling, doping and nonadiabatic effects. *Solid State Commun.* **2007**, *143*, 47–57. [CrossRef]
38. Wang, Y.Y.; Ni, Z.H.; Yu, T.; Shen, Z.X.; Wang, H.M.; Wu, Y.H.; Chen, W.; Wee, A.T.S. Raman Studies of Monolayer Graphene: The Substrate Effect. *J. Phys. Chem. C* **2008**, *112*, 10637–10640. [CrossRef]
39. Röhrl, J.; Hundhausen, M.; Emtsev, K.V.; Seyller, T.; Graupner, R.; Ley, L. Raman spectra of epitaxial graphene on SiC(0001). *Appl. Phys. Lett.* **2008**, *92*, 201918. [CrossRef]
40. Fadil, D.; Wei, W.; Deng, M.; Fregonese, S.; Strupinski, W.; Pallecchi, E.; Happy, H. 2D-Graphene Epitaxy on SiC for RF Application: Fabrication, Electrical Characterization and Noise Performance. In Proceedings of the 2018 IEEE/MTT-S International Microwave Symposium—IMS, Philadelphia, PA, USA, 10–15 June 2018; pp. 228–231.
41. Passi, V.; Gahoi, A.; Ruhkopf, J.; Kataria, S.; Vaurette, F.; Pallecchi, E.; Happy, H.; Lemme, M.C. Contact resistance Study of "edge-contacted" metal-graphene interfaces. In Proceedings of the ESSDERC 2016—46th European Solid-State Device Research Conference, Lausanne, Switzerland, 12–15 September 2016; pp. 236–239.
42. Anzi, L.; Mansouri, A.; Pedrinazzi, P.; Guerriero, E.; Fiocco, M.; Pesquera, A.; Centeno, A.; Zurutuza, A.; Behnam, A.; A Carrion, E.; et al. Ultra-low contact resistance in graphene devices at the Dirac point. *2D Mater.* **2018**, *5*, 025014. [CrossRef]
43. Watanabe, E.; Conwill, A.; Tsuya, D.; Koide, Y. Low contact resistance metals for graphene based devices. *Diam. Relat. Mater.* **2012**, *24*, 171–174. [CrossRef]
44. Das, A.; Pisana, S.; Chakraborty, B.; Piscanec, S.; Saha, S.K.; Waghmare, U.V.; Novoselov, K.S.; Krishnamurthy, H.R.; Geim, A.K.; Ferrari, A.C.; et al. Monitoring dopants by Raman scattering in an electrochemically top-gated graphene transistor. *Nat. Nanotechnol.* **2008**, *3*, 210–215. [CrossRef]
45. Bruna, M.; Ott, A.K.; Ijäs, M.; Yoon, D.; Sassi, U.; Ferrari, A.C. Doping Dependence of the Raman Spectrum of Defected Graphene. *ACS Nano* **2014**, *8*, 7432–7441. [CrossRef]
46. Ni, Y.; Ma, M.; Wei, H.; Gong, J.; Han, H.; Liu, L.; Xu, Z.; Xu, W. Multiplexed neurotransmission emulated for emotion control. *Nano Energy* **2021**, *86*, 106038. [CrossRef]

47. Asad, M.; Bonmann, M.; Yang, X.; Vorobiev, A.; Jeppson, K.; Banszerus, L.; Otto, M.; Stampfer, C.; Neumaier, D.; Stake, J. The Dependence of the High-Frequency Performance of Graphene Field-Effect Transistors on Channel Transport Properties. *IEEE J. Electron. Devices Soc.* **2020**, *8*, 457–464. [CrossRef]
48. Nougaret, L.; Happy, H.; Dambrine, G.; Derycke, V.; Bourgoin, J.-P.; Green, A.A.; Hersam, M.C. 80 GHz field-effect transistors produced using high purity semiconducting single-walled carbon nanotubes. *Appl. Phys. Lett.* **2009**, *94*, 243505. [CrossRef]

materials

Article

Al_2O_3 Layers Grown by Atomic Layer Deposition as Gate Insulator in 3C-SiC MOS Devices

Emanuela Schilirò *, Patrick Fiorenza, Raffaella Lo Nigro, Bruno Galizia, Giuseppe Greco, Salvatore Di Franco, Corrado Bongiorno, Francesco La Via, Filippo Giannazzo and Fabrizio Roccaforte

Consiglio Nazionale delle Ricerche, Istituto per la Microelettronica e Microsistemi (CNR-IMM), Strada VIII 5, Zona Industriale, 95121 Catania, Italy; patrick.fiorenza@imm.cnr.it (P.F.); raffaella.lonigro@imm.cnr.it (R.L.N.); bruno.galizia@imm.cnr.it (B.G.); giuseppe.greco@imm.cnr.it (G.G.); salvatore.difranco@imm.cnr.it (S.D.F.); corrado.bongiorno@imm.cnr.it (C.B.); francesco.lavia@imm.cnr.it (F.L.V.); filippo.giannazzo@imm.cnr.it (F.G.); fabrizio.roccaforte@imm.cnr.it (F.R.)

* Correspondence: emanuela.schiliro@imm.cnr.it

Abstract: Metal-oxide-semiconductor (MOS) capacitors with Al_2O_3 as a gate insulator are fabricated on cubic silicon carbide (3C-SiC). Al_2O_3 is deposited both by thermal and plasma-enhanced Atomic Layer Deposition (ALD) on a thermally grown 5 nm SiO_2 interlayer to improve the ALD nucleation and guarantee a better band offset with the SiC. The deposited Al_2O_3/SiO_2 stacks show lower negative shifts of the flat band voltage V_{FB} (in the range of about −3 V) compared with the conventional single SiO_2 layer (in the range of −9 V). This lower negative shift is due to the combined effect of the Al_2O_3 higher permittivity ($\varepsilon = 8$) and to the reduced amount of carbon defects generated during the short thermal oxidation process for the thin SiO_2. Moreover, the comparison between thermal and plasma-enhanced ALD suggests that this latter approach produces Al_2O_3 layers possessing better insulating behavior in terms of distribution of the leakage current breakdown. In fact, despite both possessing a breakdown voltage of 26 V, the T-ALD Al_2O_3 sample is characterised by a higher current density starting from 15 V. This can be attributable to the slightly inferior quality (in terms of density and defects) of Al_2O_3 obtained by the thermal approach and, which also explains its non-uniform dC/dV distribution arising by SCM maps.

Keywords: high-κ; dielectrics; ALD; WBG; 3C-SiC

check for updates

Citation: Schilirò, E.; Fiorenza, P.; Lo Nigro, R.; Galizia, B.; Greco, G.; Di Franco, S.; Bongiorno, C.; La Via, F.; Giannazzo, F.; Roccaforte, F. Al_2O_3 Layers Grown by Atomic Layer Deposition as Gate Insulator in 3C-SiC MOS Devices. *Materials* **2023**, *16*, 5638. https://doi.org/10.3390/ma16165638

Academic Editor: Alexander A. Lebedev

Received: 1 August 2023
Revised: 10 August 2023
Accepted: 14 August 2023
Published: 15 August 2023

1. Introduction

The cubic polytype of silicon carbide (3C-SiC) has a smaller energy gap (E_g = 2.36 eV) [1,2] compared to the hexagonal 4H-SiC (E_g = 3.26 eV) [3], but it possesses a higher electron mobility and saturation velocity [4–8]. Moreover, it exhibits a larger conduction band offset (3.7 eV) [9] with SiO_2 than 4H-SiC (2.7 eV). Hence, differently from the 4H-SiC/SiO_2 system where they are aligned with the conduction band edge of 4H-SiC, the near-interface-oxide-traps (NIOTs) inside the insulator in the 3C-SiC/SiO_2 system lie above the Fermi level and hence they are electrically inactive [10,11]. Furthermore, the lower position of the 3C-SiC conduction band edge with respect to the SiO_2 conduction band edge results immune to the interface states that are peculiar of the SiO_2/4H-SiC interface [6]. This can lead to a higher inversion electron channel mobility (>200 $cm^2\ V^{-1}\ s^{-1}$ [12]) in 3C-SiC Metal-Oxide-Semiconductor Field-Effect Transistors (MOSFETs) compared to those fabricated using the 4H-poly-type.

Silicon dioxide (SiO_2) is the native oxide of SiC that can be obtained by a thermal oxidation process of the material [13,14]. However, its electrical behavior is adversely affected by the large number of defects [9,15] (e.g., carbon clusters and dangling bonds produced during oxidation), which results in a large negative shift of the flat band voltage (V_{FB}) [8,16,17]. Another issue is the response of the MOS system to the application of high voltages. In particular, in blocking configuration, the distribution of the electric field inside the insulator (E_{ins}) and the semiconductor (E_s) can be expressed by the Gauss' law,

$E_{ins} = (\kappa_S/\kappa_{ins})E_S$, where κ_{ins} and κ_s are the insulator and semiconductor permittivity values. Considering the permittivity values of SiO_2 (3.9) and 3C-SiC (9.7), the SiO_2 layer is subjected to an electric field 2.5 higher than 3C-SiC. Hence, the SiO_2 gate insulator reliability is seriously compromised under high electric field. Moreover, using SiO_2 does not enable full exploitation of the high critical field of the underlying 3C-SiC substrate. Consequently, thicker drift layer must be used, which in turn increases the total device on-resistance [18].

Insulators with high permittivity (the so-called "*high-κ*") can be a solution to overcome this limitation due to the better distribution of the electric field in the MOS system, which offers safer operating conditions in high voltage applications. Al_2O_3 is a suitable *high-κ* oxide due to its permittivity value ($\kappa \sim 9$), good thermal stability and relatively large band gap (~7 eV) [19–23]. The Atomic Layer Deposition (ALD) [24,25] is the best technique for the deposition of Al_2O_3 thin layers with optimal thickness control, uniformity on large area, and high-quality interface [26–28]. The ALD growth of Al_2O_3 thin films on SiC can be improved by the insertion of a nanometric SiO_2 interlayer (IL), which provides a larger amount of active nucleation sites than the bare SiC surface. Moreover, the introduction of SiO_2 IL between Al_2O_3 and SiC is also convenient to guarantee a larger conduction band offset and finally to better prevent leakage phenomena [29,30]. To date, the Al_2O_3 deposited by ALD as gate dielectric on 3C-SiC is completely unexplored. Actually, it has been adopted by R. Oka et al. [16] only as a thin interlayer between SiO_2 and 3C-SiC to improve the structural quality of their interface.

In this work, we report on the Al_2O_3 thin film growth by ALD as an alternative insulator layer for 3C-SiC MOS capacitors using a very thin SiO_2 film as IL. In particular, the structural properties of Al_2O_3/SiO_2 stacks, of their interfaces on the underlying 3C-SiC but also their electrical behavior have been investigated by comparing the two different ALD approaches, namely the thermal (T-) and plasma-enhanced (PE-) ALD processes [31,32]. Both approaches allow obtaining good quality high-κ dielectrics. However, some literature works [16,22,23,31,33,34] report on slight differences both in the quality of the grown high-κ and in its interfacial properties, directly related to the different oxidation mechanism between the two methods. In particular, the studies conducted on other semiconductor materials [16,32] suggest that the more reactive action of the O_2-plasma produces Al_2O_3 layers characterised by a higher mass density and lower amounts of the undesired carbon contaminations and unreacted OH- groups, which could act as active centres for electron trapping [35].

Furthermore, the combination of several characterisation techniques, i.e., morphological-structural and electrical—either at a macroscopic scale or at a nano-scale—allowed the full comprehension of the insulating properties of the differently ALD deposited Al_2O_3 films.

2. Materials and Methods

A 10.2 μm thick 3C-SiC grown by chemical vapour deposition on Si (100) was employed as substrate [36]. Prior to the oxidation process, the 3C-SiC substrates were cleaned for ten minutes in an H_2SO_4:H_2O_2 = 3:1 solution followed by ten minutes of etching in an HF:H_2O = 1:5 solution. A 5 nm SiO_2 IL was grown by a controlled dry oxidation process at 1150 °C for 5 min. Successively, the Al_2O_3 layers were deposited on SiO_2/3C-SiC by either thermal- or plasma-enhanced ALD using trimethylaluminum (TMA) as an aluminium precursor and H_2O or O_2-plasma as co-reactants. Both processes were carried out at the deposition temperature of 250 °C. Meanwhile, the different growth rates of the T-ALD (~0.9 Å/cycle) and the PE-ALD (~1.2 Å/cycle) involve the use of a different number of deposition cycles (350 and 250 cycles for T- and PE-, respectively) to grow an A_2O_3 layer with the same thickness of 30 nm.

The structural quality of Al_2O_3/SiO_2/3C-SiC stacks and their morphology were investigated by transmission electron microscopy (TEM) using a FEG-TEM JEOL 2010F (Tokyo, Japan) microscope and by atomic force microscopy (AFM) using a DI3100 equipment by Bruker (Billerica, MA, USA) with Nanoscope V controller, respectively. In particular, the TEM analysis was carried out in cross-section in order to visualize the properties of the

Al_2O_3/SiO_2 stack layer and their interfaces. For this purpose, cross-sectional specimens were properly prepared both for T- and PE-Al_2O_3/SiO_2/3C-SiC samples by conventional mechanical preparation techniques, i.e., including polishing and dimple grinding, followed by a final thinning with ion milling.

The electrical behaviour of the insulating stacks was evaluated by capacitance-voltage (C-V) and current-voltage (I-V) measurements carried out on lateral metal-oxide-semiconductor (MOS) capacitors using a Microtech Cascade probe station equipped with a Keysight B1505 parameter analyser (Santa Rosa, CA, USA). Finally, the nanoscale electrical behaviour of the systems was monitored my means of scanning capacitance microscopy (SCM).

3. Discussion

The cross-section TEM image reported in Figure 1 illustrates a uniform and amorphous Al_2O_3 layer with a thickness of ~30 nm and a sharp interface with the underlying SiO_2/3C-SiC. The SiO_2-IL is clearly distinguishable and has a thickness of about 4.5 nm. The structural properties of Al_2O_3 and of its interfaces are similar on both T-Al_2O_3/SiO_2/3C-SiC and PE-Al_2O_3/SiO_2/3C-SiC systems; thus, only the first is reported representatively.

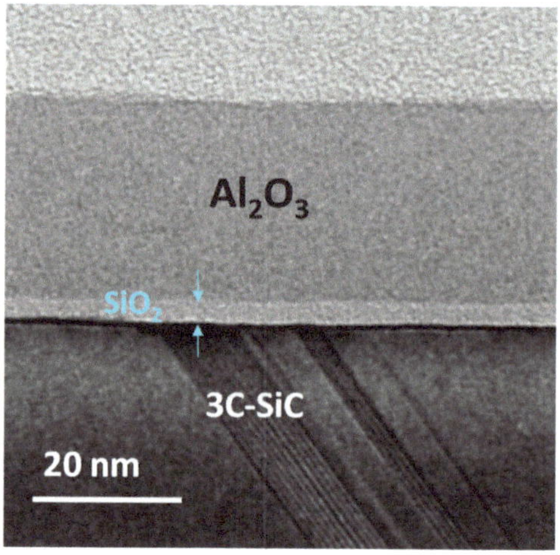

Figure 1. Cross-section TEM image relative to T-Al_2O_3/SiO_2/3C-SiC.

Lateral MOS capacitors schematically depicted in Figure 2a were fabricated using photolithography, metal deposition and lift-off processes. The anode of the MOS capacitors was surrounded by a large-area metal cathode so that its capacitance could be neglected (Figure 2b). Ni/Au was used as metal electrode. The MOS structures fabricated on T-Al_2O_3/SiO_2/3C-SiC and PE-Al_2O_3/SiO_2/3C-SiC were probed by C-V measurements, which are shown in Figure 2c. Both samples provide C-V curves negatively shifted compared to the ideal value V_{FB} = +0.9 V. In particular, the experimental flat band voltage values were −0.6 V and −3 V for the T-ALD and PE-ALD stacks, respectively. However, as can be observed in Figure 2c, such negative shifts were smaller compared to that of MOS capacitor where the insulator was only a thick (40 nm) thermal SiO_2 [37]. This experimental finding is related to the higher dielectric constant of the Al_2O_3 (κ = 8) with respect to that of SiO_2 (κ = 3.9). In fact, even though the SiO_2/3C-SiC interface is similar, in both cases resulting in analogous amount of effective charge (N_{eff}), this can cause a variation of the experimental V_{FB} value, moving it toward the ideal one, as expressed by the following equations:

$$\Delta V_{FB} = \frac{qN_{eff}}{C_{OX}}, \quad (1)$$

$$C_{OX} = \frac{\varepsilon_0\ \kappa}{t_{OX}}, \quad (2)$$

where N_{eff} is the effective trapped charge density, C_{OX} is the accumulation capacitance, q is the electron charge, ε_0 is the vacuum dielectric constant, and t_{OX} is the oxide thickness.

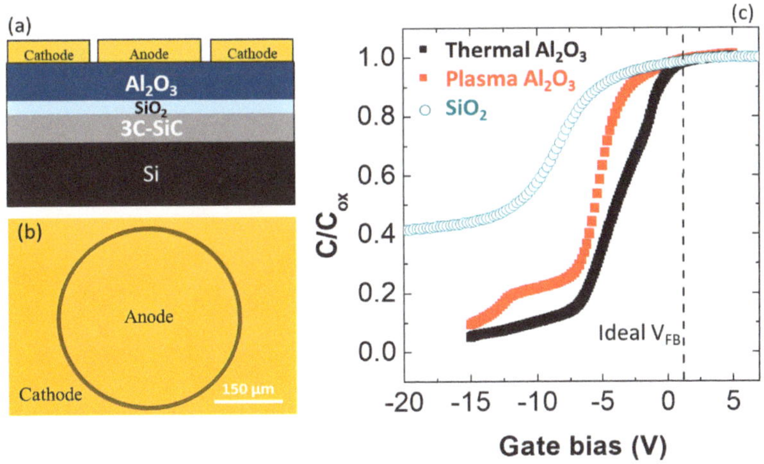

Figure 2. Schematic cross-section of the Al_2O_3/SiO_2/3C-SiC MOS capacitor (**a**) and top-view microscopy image of a MOS capacitor (**b**). C-V curves of T-Al_2O_3/SiO_2(IL)/3C-SiC and PE-Al_2O_3/SiO_2(IL)/3C-SiC MOS capacitors in comparison with the analogous SiO_2/3C-SiC (**c**).

According to Equations (1) and (2), for a constant N_{eff} and an insulating layer thickness, an increased dielectric constant results in a smaller flat band voltage shift ΔV_{FB}. Furthermore, the lower negative V_{FB} shift of the Al_2O_3/SiO_2/3C-SiC stack can be also explained by the shorter time for the thermally oxidation process needed to grow a 4.5 nm SiO_2 IL than that needed to grow a 30 nm thick SiO_2. In fact, a shorter oxidation time produces a lower amount of carbon clusters responsible of the negative V_{FB} shift [38]. From the accumulation capacitance, the dielectric constant κ of the insulating films was estimated to be ~8 both for T- and PE-Al_2O_3. As can be seen in Figure 2b, the C-V curves of the PE- and T-ALD Al_2O_3/SiO_2/3C-SiC are characterised by a different electrical behavior. In fact, besides the negative flat band voltage shift occurring in both cases, it can be noticed that a bump was visible in the depletion region of the C-V curve of the PE-ALD sample. Plausibly, this bump was caused by the occurrence of charge trapping at deep interface states when increasing the bias [12]. On the other hand, the thermal Al_2O_3/SiO_2/3C-SiC sample is characterised by a more pronounced stretch-out of the C-V curve. Evidently, the different nature of the oxidation process (plasma enhanced-PE, and thermal-T) used during the ALD growth of Al_2O_3 was responsible for the different electrical quality of the two interfaces.

From the C-V curves, by applying the Terman's method [39], the interface states (D_{it}) distributions were calculated for both T-Al_2O_3/SiO_2/3C-SiC and PE-Al_2O_3/SiO_2/3C-SiC stacks, which are also reported in Figure 3 in comparison to that of the SiO_2/3C-SiC system. The SiO_2/3C-SiC and Al_2O_3/SiO_2/3C-SiC samples exhibited a comparable D_{it} distribution in the order of 2×10^{12} cm^{-2} eV^{-1} [8]. On the other hand, the PE-Al_2O_3/SiO_2/3C-SiC sample showed a lower D_{it} distribution close to the 3C-SiC conduction band edge in the order of 5×10^{11} cm^{-2} eV^{-1}, which can be due to the beneficial effect of the O_2-plasma on the defects amount at SiO_2/SiC interface [40]. In fact, Kim et al. [30] demonstrated that for the SiO_2/SiC-based devices, the use of a SiO_2 growth process assisted by the highly reactive O_2 plasma guarantees the formation of an interface characterised by a lower amount of defects and more stable SiO bonds. Analogously, in our case, the PE-approach used to deposit the Al_2O_3 could play a similar beneficial effect on the underlying SiO_2/SiC interface.

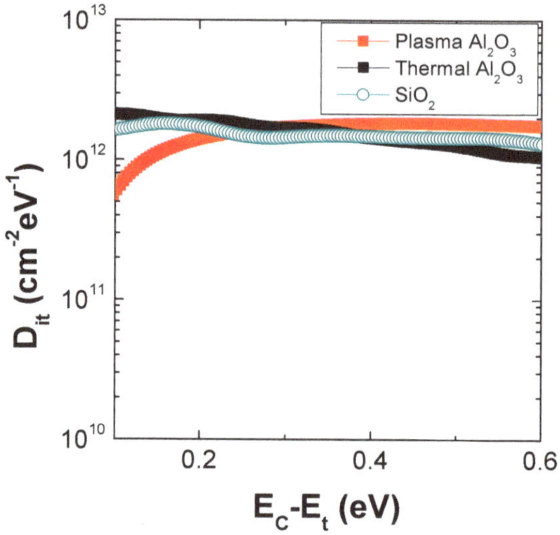

Figure 3. D_{it} distribution of T-Al_2O_3/SiO_2(IL)/3C-SiC and PE-Al_2O_3/SiO_2(IL)/3C-SiC MOS capacitors in comparison with the analogous SiO_2/3C-SiC.

The current–voltage (I-V) curves acquired on the T-Al_2O_3/SiO_2/3C-SiC and PE-Al_2O_3/SiO_2/3C-SiC MOS capacitors are shown in Figure 4. As can be seen, in both systems, the electrical breakdown occurred at a gate bias of over 26 V. However, while the PE-Al_2O_3/SiO_2/3C-SiC sample maintained a constant current value of 10^{-12} A up to the breakdown, the T-Al_2O_3/SiO_2/3C-SiC sample exhibited a fast raise of the current starting from 15 V. The leakage current trend occurring across the Al_2O_3 layer deposited by thermal mode could be explained by a slightly lower mass density and a higher amount of -OH and/or -CH_3 groups than that deposited by the plasma-enhanced mode due to the less efficacious oxidation process by the H_2O-precursor [41–43]. Moreover, in comparison to the I-V curve typical of the 3C-SiC capacitor with a 40 nm thick SiO_2 as a dielectric layer (also reported in Figure 4), which exhibited a breakdown voltage of about 20 V, both T-Al_2O_3/SiO_2 and PE-Al_2O_3/SiO_2 stacks were able to shift the breakdown phenomena toward higher voltages, over 26 V. The early breakdown of a thick thermal grown SiO_2 on 3C-SiC has already been explained by F. Li et al. [44] as a consequence of the large amount of carbon left during the thermal oxidation process. However, in our case, the use of a short oxidation process to obtain only a thin IL probably resulted in a smaller amount of carbon defects to cause the early breakdown.

The electrical behavior of both T- and PE-Al_2O_3/SiO_2/3C-SiC stacks was studied at the nanoscale by SCM measurement. A schematic representation of the SCM experimental setup is illustrated in Figure 5a. During the surface scan with a diamond tip, an AC modulating bias at 100 kHz frequency and with amplitude $\Delta V = 2$ V (below the conduction regime through the insulator) was applied to the sample, and the capacitance variation ΔC in response to this modulation was recorded with the SCM sensor. Figure 5b,c show the AFM morphology of the T- and PE-Al_2O_3/SiO_2/3C-SiC stacks on the portion of the samples where the SCM maps were acquired. The highly irregular morphology is peculiar of the 3C-SiC material [18], which is characterised by terraces separated by anti-phase boundaries. The SCM maps of T- and PE-samples are reported in Figure 5d,e. The SCM signal is a result of the capacitance change (dC/dV) in the local metal-insulator-semiconductor capacitor, where the metal is the conductive AFM tip. Hence, the SCM response depends on the semiconductor characteristics (i.e., doping type and concentration) but also on the insulator properties (including thickness, interface state density, oxide traps, and permittivity) [45,46]. Considering that both the T- and PE-Al_2O_3 layers were deposited on the same 3C-SiC substrate and that they are characterised by an equivalent interface with SiO_2 as IL, the different SCM maps (Figure 5d,e) obtained for the two cases can be correlated to the different insulator quality. In particular, the SCM map of the T-Al_2O_3/SiO_2/3C-SiC stack, reported in Figure 5d, shows a non-uniform dC/dV signal distribution visible as the change in the color gradient from one spot to another. In contrast, the PE- Al_2O_3/SiO_2/3C-SiC stack reported in Figure 5e exhibits a well-uniform SCM map, with only a small deviating region. The different SCM responses between T- and PE-systems could be due to the different structural quality of the Al_2O_3 layers deposited by the two approaches. In fact, the different Al_2O_3 quality, in terms of mass density and/or -OH/-CH_3 contaminations, which can arise by using the different oxidation processes (T- or PE-), determines its charge trapping behavior and permittivity and, ultimately, the SCM signal. Similar results have been previously reported for the growth of Al_2O_3 thin layers on AlGaN/GaN heterostructures by the two different T-ALD and PE-ALD approaches, where the evolution of the insulating behavior investigated at the nanoscale upon increasing film thickness clearly indicated a different nucleation mechanism [16]. Hence, the present investigation at the nano-scale also confirms the better electrical performance of the PE-Al_2O_3 layer already observed by the electrical measurements acquired on the macroscopic capacitors.

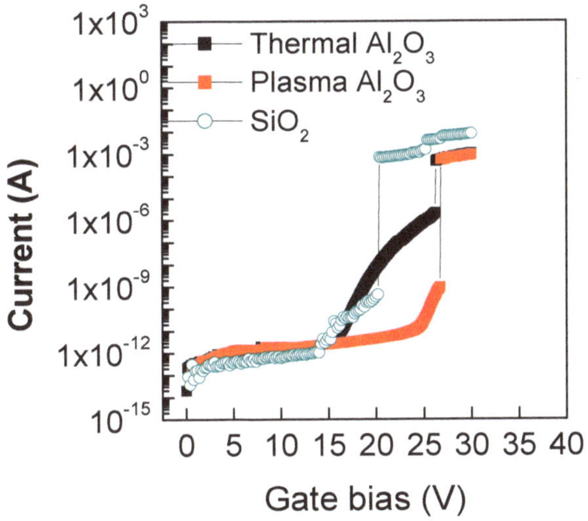

Figure 4. I-V measurements of T-Al_2O_3/SiO_2/3C-SiC and PE-Al_2O_3/SiO_2/3C-SiC MOS capacitors in comparison with the analogous SiO_2/3C-SiC.

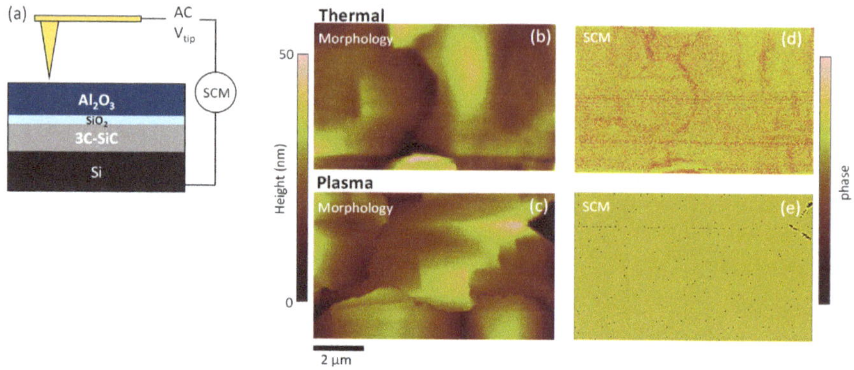

Figure 5. Schematic of the SCM experimental setup (**a**). AFM morphology of T-Al_2O_3/SiO_2/3C-SiC (**b**) and PE-Al_2O_3/SiO_2/3C-SiC (**c**). SCM maps of T-Al_2O_3/SiO_2/3C-SiC (**d**) and PE-Al_2O_3/SiO_2/3C-SiC (**e**).

4. Conclusions

The insulating properties of the Al_2O_3 layers deposited on 3C-SiC both by the thermal- and plasma-enhanced ALD approaches were investigated. Our results demonstrated that:

- A thin (5 nm) SiO_2 IL between the Al_2O_3 and the 3C-SiC is useful to ensure the quality of ALD growth and to maximize the insulator/semiconductor band offset;
- The Al_2O_3 is a valid alternative to the conventional thermally grown single SiO_2 as gate insulator for 3C-SiC MOS-based devices. In fact, the Al_2O_3 layers showed a high permittivity (~8), which produced a significant reduction in the negative flat band voltage shift that is usually observed with SiO_2;
- A different electrical behavior was found between thermal- and plasma-enhanced Al_2O_3 both by investigations on macroscopic MOS capacitors and at the nanoscale using SCM analysis. In fact, although both systems ensure an electrical breakdown over 26 V, the T-Al_2O_3/SiO_2/3C-SiC stack exhibits early leakage phenomena already from 15 V. Moreover, the T-Al_2O_3/SiO_2/3C-SiC is characterised by a non-uniform SCM map compared to the PE-Al_2O_3/SiO_2/3C-SiC. This difference can be correlated to a different Al_2O_3 quality obtained through the two different oxidation processes (T- or PE-), resulting in an inhomogeneous charge trapping behavior and permittivity.

These results can be important for the fabrication of 3C-SiC MOSFETs with a positive turn-on voltage with improved channel conduction properties.

Author Contributions: Conceptualisation, E.S., P.F., R.L.N. and F.R.; methodology, E.S., P.F., R.L.N., S.D.F., G.G., F.L.V., F.G. and F.R.; validation, P.F., R.L.N. and F.R.; formal analysis, E.S. and P.F.; investigation, E.S., P.F., B.G. and C.B.; data curation, E.S. and P.F.; writing—original draft preparation, E.S.; writing—review and editing, P.F., R.L.N. and F.R.; supervision, F.R.; project administration, F.R.; funding acquisition, F.R. and F.L.V. All authors have read and agreed to the published version of the manuscript.

Funding: This work has been partially supported by the European project CHALLENGE (Grant Agreement 720827). Moreover, the research received funding from the European Union (NextGeneration EU), through the MUR-PNRR projects SAMOTHRACE (ECS00000022) and iEntrance@ENL (IR0000027). Part of the experiments reported in this paper have been carried out in the Italian Infrastructure Beyond-Nano.

Institutional Review Board Statement: Not applicable.

Informed Consent Statement: Not applicable.

Data Availability Statement: Data is contained within the article.

Acknowledgments: The authors would like to acknowledge M. Vivona and S. Panasci for fruitful discussions on the results.

Conflicts of Interest: The authors declare no conflict of interest.

Abbreviations

ALD	Atomic Layer Deposition
T-ALD	Thermal-ALD
PE-ALD	Plasma Enanched-ALD
NIOTs	Near Interface Oxide Traps
MOSFET	Metal Oxide Semiconductor Field Effect Transistor
MOS	Metal Oxide Semiconductor
IL	Interlayer
High-κ	High permittivity (κ) dielectrics
TEM	Transmission Electron Microscopy
AFM	Atomic Force Microscopy
SCM	Scanning Capacitance Microscopy

References

1. Levinshein, M.; Sergey, L.; Shur, M. *Properties of Advanced Semiconductor Materials: GaN, AIN, InN, BN, SiC, SiGe*, 1st ed.; John Wiley & Sons, Inc.: New York, NY, USA, 2001.
2. Bimberg, D.; Altarelli, M.; Lipari, N. A calculation of valence band masses, exciton and acceptor energies and the ground state properties of the electron-hole liquid in cubic SiC. *Solid State Commun.* **1981**, *40*, 437–440. [CrossRef]
3. Itoh, A.; Akita, H.; Kimoto, T.; Matsunami, H. High quality 4H-SiC homoepitaxial layers grown by step controlled epitaxy. *Appl. Phys. Lett.* **1994**, *65*, 1400–1402. [CrossRef]
4. Arvanitopoulos, A.E.; Antoniou, M.; Perkins, S.; Jennings, M.; Guadas, M.B.; Gyftakis, K.N.; Lophitis, N. On the Suitability of 3C-Silicon Carbide as an Alternative to 4H-Silicon Carbide for Power Diodes. *IEEE Trans. Ind. Appl.* **2019**, *55*, 4080–4090. [CrossRef]
5. La Via, F.; Severino, A.; Anzalone, R.; Bongiorno, C.; Litrico, G.; Mauceri, M.; Schoeler, M.; Schuh, P.; Wellmann, P. From thin film to bulk 3C-SiC growth: Understanding the mechanism of defects reduction. *Mater. Sci. Semicond. Process.* **2018**, *78*, 57–68. [CrossRef]
6. Roccaforte, F.; Greco, G.; Fiorenza, P.; Di Franco, S.; Giannazzo, F.; La Via, F.; Zielinski, M.; Mank, H.; Jokubavicius, V.; Yakimova, R. Towards vertical Schottky diodes on bulk cubic silicon carbide (3C-SiC). *Appl. Surf. Sci.* **2022**, *606*, 154896. [CrossRef]
7. Giannazzo, F.; Greco, G.; Di Franco, S.; Fiorenza, P.; Deretzis, I.; La Magna, A.; Bongiorno, C.; Zimbone, M.; La Via, F.; Zielinski, M.; et al. Impact of Stacking Faults and Domain Boundaries on the Electronic Transport in Cubic Silicon Carbide Probed by Conductive Atomic Force Microscopy. *Adv. Electron. Mater.* **2020**, *6*, 1901171. [CrossRef]
8. Renz, A.B.; Li, F.; Vavasour, O.J.; Gammon, P.M.; Dai, T.; Baker, G.W.C.; La Via, F.; Zielinski, M.; Zhang, L.; Grant, N.E.; et al. Initial investigations into the MOS interface of freestanding 3C-SiC layers for device applications. *Semicond. Sci. Technol.* **2021**, *36*, 055006. [CrossRef]
9. Afanasev, V.V.; Bassler, M.; Pensl, G.; Schulz, M. Intrinsic SiC/SiO_2 Interface States. *Phys. Status Solidi (A)* **1997**, *162*, 321–337. [CrossRef]
10. Li, F.; Roccaforte, F.; Greco, G.; Fiorenza, P.; La Via, F.; Pérez-Tomas, A.; Evans, J.E.; Fisher, C.A.; Monaghan, F.A.; Mawby, P.A.; et al. Status and Prospects of Cubic Silicon Carbide Power Electronics Device Technology. *Materials* **2021**, *14*, 5831. [CrossRef]
11. Esteve, R. Fabrication and Characterization of 3C- and 4H-SiC MOSFETs. Ph.D. Thesis, KTH Royal Institute of Technology, School of Information and Communication Technology (ICT), Integrated Devices and Circuits, Stockholm, Sweden, 2011.
12. Lee, K.K.; Ishida, Y.; Ohshima, T.; Kojima, K.; Tanaka, Y.; Takahashi, T.; Okumura, H.; Arai, K.; Kamiya, T. N-Channel MOSFETs Fabricated on Homoepitaxy-Grown 3C-SiC Films. *IEEE Electron Device Lett.* **2003**, *24*, 466–468.
13. Roccaforte, F.; Fiorenza, P.; Greco, G.; Vivona, M.; Nigro, R.L.; Giannazzo, F.; Patti, A.; Saggio, M. Recent advances on dielectrics technology for SiC and GaN power devices. *Appl. Surf. Sci.* **2014**, *301*, 9–18. [CrossRef]
14. Lophitis, N.; Arvanitopoulos, A.; Jennings, M.R.; Mawby, P.A.; Antoniou, M. On the 3C-SiC/SiO_2 n-MOS interface and the creation of a calibrated model for the Electrons' Inversion Layer Mobility covering a wide range of operating temperatures and applied gate voltage. In Proceedings of the IEEE Workshop on Wide Bandgap Power Devices and Applications in Europe (WiPDA Europe), Coventry, UK, 18–20 September 2022.
15. Ciobanu, F.; Pensl, G.; Nagasawa, H.; Schöner, A.; Dimitrijev, S.; Cheong, K.Y.; Afanas'ev, V.V.; Wagner, G. Traps at the Interface of 3C-SiC/SiO_2-MOS-Structures. In *Materials Science Forum*; Trans Tech Publications Ltd.: Zurich, Switzerland, 2003; Volume 433, pp. 551–554.

16. Oka, R.; Yamamoto, K.; Akamine, H.; Wang, D.; Nakashima, H.; Hishiki, S.; Kawamura, K. High interfacial quality metal-oxide-semiconductor capacitor on (111) oriented 3C-SiC with Al_2O_3 interlayer and its internal charge analysis. *Jpn. J. Appl. Phys.* **2020**, *59*, SGGD17. [CrossRef]
17. Cherkaoui, K.; Blake, A.; Gomeniuk, Y.Y.; Lin, J.; Sheehan, B.; White, M.; Hurley, P.K.; Ward, P.J. Investigating positive oxide charge in the SiO_2/3C-SiC MOS system. *AIP Adv.* **2018**, *8*, 085323. [CrossRef]
18. Baliga, B.J. *Silicon Carbide Power Devices*; World Scientific Publishing Co. Pte. Ltd.: Singapore, 2005.
19. Lo Nigro, R.; Fiorenza, P.; Greco, G.; Schilirò, E.; Roccaforte, F. Structural and Insulating Behaviour of High-Permittivity Binary Oxide Thin Films for Silicon Carbide and Gallium Nitride Electronic Devices. *Materials* **2022**, *15*, 830. [CrossRef]
20. Gao, K.Y.; Seyller, T.; Ley, L.; Ciobanu, F.; Pensl, G.; Tadich, A.; Riley, J.D.; Leckey, R.G.C. Al_2O_3 prepared by atomic layer deposition as gate dielectric on 6H-SiC(0001). *Appl. Phys. Lett.* **2003**, *83*, 1830. [CrossRef]
21. Yu, Y.; Jun, H.; Yun, D.Y.; An, K.; Zhan, C.; Yu, X. Influences of high-temperature annealing on atomic layer deposited Al2O3/4H-SiC. *Chin. Phys. B* **2013**, *22*, 07810.
22. Lin, H.C.; Ye, P.D.; Wilk, G.D. Leakage current and breakdown electric-field studies on ultrathin atomic-layer-deposited Al_2O_3 on GaAs. *Appl. Phys. Lett.* **2005**, *87*, 182904. [CrossRef]
23. Avice, M.; Grossner, U.; Pintilie, I.; Svesson, G.; Servidori, M.; Nipoti, R.; Nilsen, O.; Fjellvag, H. Low Density of Near-Interface Traps at the Al_2O_3/4H-SiC Interface with Al_2O_3 Made by Low Temperature Oxidation of Al. *Mater. Sci. Forum* **2007**, *897*, 135–138.
24. Kääriäinen, T.; Cameron, D.; Kääriäinen, M.-L.; Sherman, A. *Atomic Layer Deposition, Principles Characteristics and Nanotechnolody Applications*; Scrivener, M., Carmical, P., Eds.; John Wiley & Sons: Hoboken, NJ, USA, 2013.
25. Johnson, R.W.; Hultqvist, A.; Bent, S.F. A brief review of atomic layer deposition: From fundamentals to applications. *Mater. Today* **2014**, *17*, 236–246. [CrossRef]
26. George, S.M. Atomic Layer Deposition: An Overview. *Chem. Rev.* **2009**, *110*, 111–131. [CrossRef]
27. Niinistö, L.; Päiväsaari, J.; Niinistö, J.; Putkonen, M.; Nieminen, M. Advanced electronic and optoelectronic materials by Atomic Layer Deposition: An overview with special emphasis on recent progress in processing of high-k dielectrics and other oxide materials. *Phys. Status Solidi (A)* **2004**, *201*, 1443–1452. [CrossRef]
28. Oviroh, P.O.; Akbarzadeh, R.; Pan, D.; Coetzee, R.A.M.; Jen, T.-C. New development of atomic layer deposition: Processes, methods and applications. *Sci. Technol. Adv. Mater.* **2019**, *20*, 465–496. [CrossRef] [PubMed]
29. Cheong, K.Y.; Moon, J.H.; Eom, D.; Kim, H.J.; Bahng, W.; Kim, N.-K. Electronic Properties of Atomic-Layer-Deposited Al_2O_3/Thermal-Nitrided SiO_2 Stacking Dielectric on 4H SiC. *Electrochem. Solid-State Lett.* **2007**, *10*, H69–H71. [CrossRef]
30. Schilirò, E.; Lo Nigro, R.; Fiorenza, P.; Roccaforte, F. Negative charge trapping effects in Al_2O_3 films grown by atomic layer deposition onto thermally oxidized 4H-SiC. *AIP Adv.* **2016**, *6*, 075021. [CrossRef]
31. van Hemmen, J.L.; Heil, S.B.S.; Klootwijk, J.H.; Roozeboom, F.; Hodson, C.J.; Van de Sanden, M.C.M.; Kessels, W.M.M. Plasma and thermal ALD of Al_2O_3 in a commercial 200 mm ALD reactor. *J. Electrochem. Soc.* **2007**, *154*, G165–G169. [CrossRef]
32. Dingemans, G.; Seguin, R.; Engelhart, P.; van de Sanden, M.C.M.; Kessels, W.M.M. Silicon surface passivation by ultrathin Al_2O_3films synthesized by thermal and plasma atomic layer deposition. *Phys. Status Solidi (RRL) Rapid Res. Lett.* **2010**, *4*, 10–12. [CrossRef]
33. Profijt, H.B.; Potts, S.E.; Van De Sanden, M.C.M.; Kessels, W.M.M. Plasma-Assisted Atomic Layer Deposition: Basics, Opportunities, and Challenges. *J. Vac. Sci. Technol. A* **2011**, *29*, 050801. [CrossRef]
34. Hoex, B.; Schmidt, J.; Pohl, P.; van de Sanden, M.C.M.; Kessels, W.M.M. On the c-Si surface passivation mechanism by the negative-charge-dielectric Al_2O_3. *J. Appl. Phys.* **2008**, *104*, 044903. [CrossRef]
35. Kotomin, E.; Popov, A. Radiation-induced point defects in simple oxides. *Nucl. Instrum. Methods Phys. Res. Sect. B Beam Interact. Mater. At.* **1998**, *141*, 1–15. [CrossRef]
36. Anzalone, R.; Privitera, S.; Camarda, M.; Alberti, A.; Mannino, G.; Fiorenza, P.; Di Franco, S.; La Via, F. Interface state density evaluation of high quality hetero-epitaxial 3C–SiC(001) for high-power MOSFET applications. *Mater. Sci. Eng. B* **2015**, *198*, 14–19. [CrossRef]
37. Fiorenza, P.; Schilirò, E.; Giannazzo, F.; Bongiorno, C.; Zielinski, M.; La Via, F.; Roccaforte, F. On the origin of the premature breakdown of thermal oxide on 3C-SiC probed by electrical scanning probe microscopy. *Appl. Surf. Sci.* **2020**, *526*, 146656. [CrossRef]
38. Newsome, D.A.; Sengupta, D.; Foroutan, H.; Russo, M.F.; van Duin, A.C.T. Oxidation of Silicon Carbide by O_2 and H_2O: A ReaxFF Reactive Molecular Dynamics Study, Part I. *J. Phys. Chem. C* **2012**, *116*, 16111–16121. [CrossRef]
39. Sze, S.M. *Physics of Semiconductor Devices*; Wiley-Interscience: Hoboken, NJ, USA, 1981; pp. 849–850.
40. Kim, D.K.; Jeong, K.S.; Kang, Y.S.; Kang, H.-K.; Cho, S.W.; Kim, S.-O.; Suh, D.; Kim, S.; Cho, M.-H. Controlling the defects and transition layer in SiO_2 films grown on 4H-SiC via direct plasma-assisted oxidation. *Sci. Rep.* **2016**, *6*, 34945. [CrossRef] [PubMed]
41. Jinesh, K.B.; van Hemmen, J.L.; van de Sanden, M.C.M.; Roozeboom, F.; Klootwijk, J.H.; Besling, W.F.A.; Kessels, W.M.M. Dielectric Properties of Thermal and Plasma-Assisted Atomic Layer Deposited Al_2O_3 Thin Films. *J. Electrochem. Soc.* **2011**, *158*, G21–G26. [CrossRef]
42. Haeberle, J.; Henkel, K.; Gargouri, H.; Naumann, F.; Gruska, B.; Arens, M.; Tallarida, M.; Schmeißer, D. Ellipsometry and XPS comparative studies of thermal and plasma enhanced atomic layer deposited Al_2O_3-films. *Beilstein J. Nanotechnol.* **2013**, *4*, 732–742. [CrossRef]

43. Schilirò, E.; Fiorenza, P.; Greco, G.; Monforte, F.; Condorelli, G.G.; Roccaforte, F.; Giannazzo, F.; Nigro, R.L. Early Growth Stages of Aluminum Oxide (Al_2O_3) Insulating Layers by Thermal- and Plasma-Enhanced Atomic Layer Deposition on AlGaN/GaN Heterostructures. *ACS Appl. Electron. Mater.* **2021**, *4*, 406–415. [CrossRef]
44. Li, F.; Qiu, S.; Jennings, M.R.; Mawby, P.A. Fabrication and Dielectric Breakdown of 3C-SiC/SiO_2 MOS Capacitors. In Proceedings of the 2019 IEEE 12th International Symposium on Diagnostics for Electrical Machines, Power Electronics and Drives (SDEMPED), Toulouse, France, 27–30 August 2019; pp. 344–350.
45. Yanev, V.; Rommel, M.; Bauer, A.J.; Frey, L. Characterization of thickness variations of thin dielectric layers at the nanoscale using scanning capacitance microscopy. *J. Vac. Sci. Technol. B* **2011**, *29*, 01A401. [CrossRef]
46. Goghero, D.; Raineri, V.; Giannazzo, F. Study of interface states and oxide quality to avoid contrast reversal in scanning capacitance microscopy. *Appl. Phys. Lett.* **2002**, *81*, 1824–1826. [CrossRef]

materials

Communication

A New Cell Topology for 4H-SiC Planar Power MOSFETs for High-Frequency Switching

Shengnan Zhu [1,*], Tianshi Liu [1], Junchong Fan [1], Arash Salemi [2], Marvin H. White [1], David Sheridan [2] and Anant K. Agarwal [1]

[1] Department of Electrical and Computer Engineering, The Ohio State University, Columbus, OH 43210, USA
[2] Alpha and Omega Semiconductor, Sunnyvale, CA 94085, USA
* Correspondence: zhu.2670@osu.edu

Abstract: A new cell topology named the dodecagonal (a polygon with twelve sides, short for Dod) cell is proposed to optimize the gate-to-drain capacitance (C_{gd}) and reduce the specific ON-resistance ($R_{on,sp}$) of 4H-SiC planar power MOSFETs. The Dod and the octagonal (Oct) cells are used in the layout design of the 650 V SiC MOSFETs in this work. The experimental results confirm that the Dod-cell MOSFET achieves a 2.2× lower $R_{on,sp}$, 2.1× smaller high-frequency figure of merit (HF-FOM), higher turn on/off dv/dt, and 29% less switching loss than the fabricated Oct-cell MOSFET. The results demonstrate that the Dod cell is an attractive candidate for high-frequency power applications.

Keywords: SiC power MOSFET; cell topology; dodecagonal cell; octagonal cell; gate-to-drain capacitance (C_{gd}); specific ON-resistance ($R_{on,sp}$); high-frequency figure of merit (HF-FOM); switching performance

check for updates

Citation: Zhu, S.; Liu, T.; Fan, J.; Salemi, A.; White, M.H.; Sheridan, D.; Agarwal, A.K. A New Cell Topology for 4H-SiC Planar Power MOSFETs for High-Frequency Switching. *Materials* **2022**, *15*, 6690. https://doi.org/10.3390/ma15196690

Academic Editor: Mike Jennings and Marilena Vivona

Received: 21 August 2022
Accepted: 21 September 2022
Published: 27 September 2022

1. Introduction

The silicon carbide (SiC) power MOSFET has been expected to be the alternative to silicon power devices in multiple applications, such as renewable energy, drivers for electrical machines, and power converters for hybrid and electric vehicles, due to its various advantages, including a low power loss, high switching frequency, and high operating temperature [1–3]. Improving energy efficiency (reducing power dissipation) is the driving force for the commercialization of SiC power MOSFETs [3]. The total power loss of a SiC power MOSFET consists of an on-state conduction loss and switching loss. The conduction loss can be reduced by reducing the ON-resistance (R_{on}). The switching loss of a SiC power MOSFET can be reduced by decreasing device capacitances [4,5]. Studies show that the layout topology design affects the on-state and dynamic performances of SiC power devices [6,7]. Different cell topologies (Linear, Hexagonal, Square, and Octagonal) were used on 600 V SiC planar MOSFETs [7]. All the figure of merits (FOMs) for the Octagonal cell topology are comparable to those of the state-of-the-art Si COOLMOS and are much better than the ROHM SiC MOSFET product [7]. In addition, simulations show that the oxide electric field for an Octagonal cell is much lower in comparison to other topologies [7]. The Linear, Hexagonal, and Octagonal cell topologies were used on 650 V SiC planar JBSFETs and reported in [8]. It is demonstrated that the Hexagonal cell topology provides the lowest specific ON-resistance ($R_{on,sp}$) but a higher gate–drain charge than Linear and Octagonal cells [8]. It indicates that the Hexagonal cell is suitable for power-switching applications at a nominal frequency. However, for applications where a faster switching frequency is desired, such as reducing the size of power converters, the layout, such as the Octagonal (Oct) cell, is preferred, which tends to reduce the C_{gd} at the cost of a higher R_{on} [9].

In this work, a new cell topology named the dodecagonal (Dod) cell is proposed and compared with the Oct cell. Additionally, 650 V SiC power MOSFETs with the Dod and Oct cells have been designed, fabricated, and characterized. The static and dynamic performances of the fabricated 650 V SiC power MOSFETs are compared.

2. Device Design and Fabrication

Figure 1a shows the structure of the proposed Dod cell. A twelve-sided P^+ region with the ohmic contact on top is surrounded by six hexagonal poly-Si gate regions. The hexagonal gate regions are connected by poly-Si bars. The hexagonal JFET regions are placed inside gate regions. This Dod cell is applied to 650 V SiC power MOSFETs. The Oct cell is shown in Figure 1b. The cross-sectional views along the AA′ direction for both Dod-cell and Oct-cell MOSFETs in this work are shown in Figure 1c. All MOSFETs have the same edge termination design and die size. The die size is 1.15×1.15 mm^2, including the termination. The MOSFETs are fabricated on a 6-inch SiC wafer by X-Fab using the same SiC power MOSFET process. Figure 1d shows the cross-sectional SEM image of the fabricated Dod-cell MOSFETs. Due to the lateral straggle of Aluminum implantation in the P-well, the narrowest portion of the JFET region is reduced by 0.2 μm on each side. Five Dod-cell and five Oct-cell MOSFETs are packaged into open-cavity TO-247 packages for device characterization. Design parameters and experimental results are listed in Table 1.

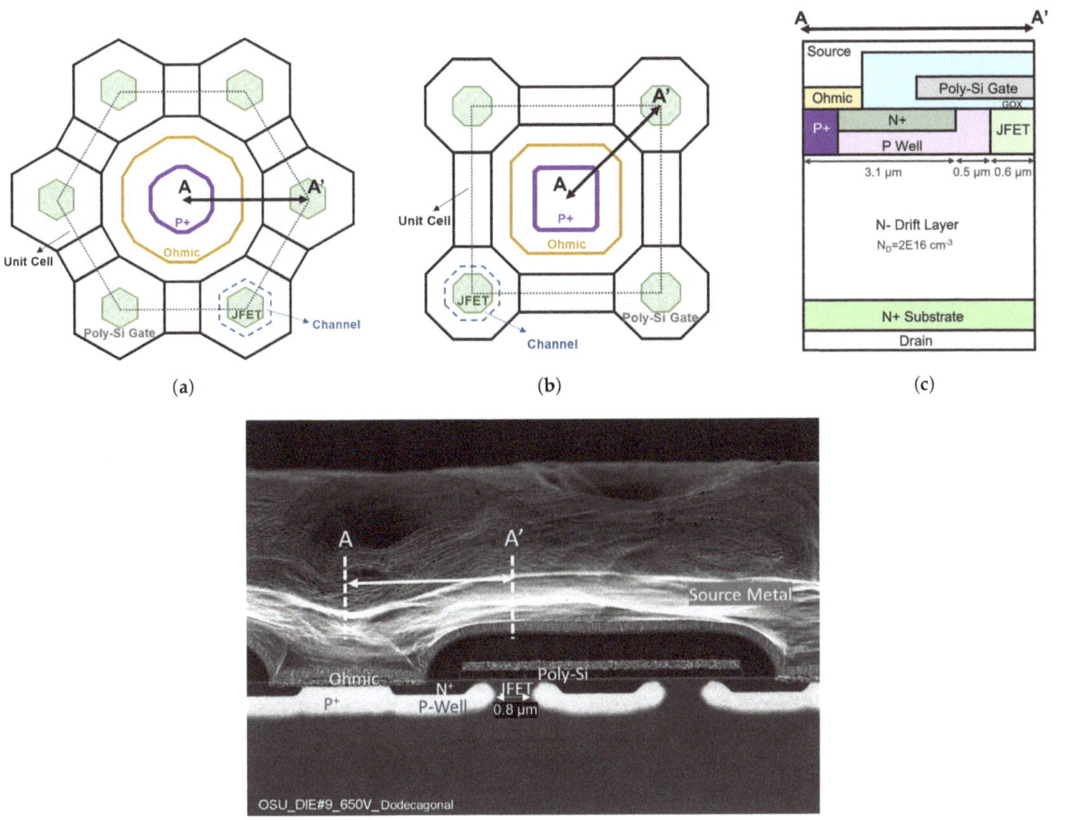

Figure 1. (**a**) Dodecagonal cell topology, (**b**) Octagonal cell topology, (**c**) A-A′ cross-sectional view of the 650 V SiC power MOSFET. (**d**) Cross-sectional SEM image of the fabricated 650 V SiC power MOSFET with the Dod cell.

Table 1. Design parameters and experimental results.

	Cell Topology	Oct	Dod
Design parameters	Cell pitch (μm)	8.4	8.4
	Active area (mm^2)	0.643	0.634
	Channel density (μm^{-1})	0.113	0.181
	JFET density	0.034	0.055
Experimental results	V_{th} (V)	4.09 ± 0.09 *	4.35 ± 0.10
	BV (V) @ I_D = 100 μA	756.3 ± 22.6	753.4 ± 23.6
	$R_{on,sp}$ (mΩ·cm^2) @ V_D = 1.5 V	24.4 ± 7.52	10.9 ± 3.06
	C_{gd} (pF) @ V_D = 400 V	1.59 ± 0.01	1.63 ± 0.04
	HF-FOM (mΩ·pF) ($R_{on} \cdot C_{gd}$)	6031 ± 1896	2817 ± 749
	dv/dt turn-on (V/ns)	6.1	12.5
	Switching loss turn-on (μJ)	29.4	20.1
	dv/dt turn-off (V/ns)	8.4	10.7
	Switching loss turn-off (μJ)	5.7	4.7
	Total switching loss (μJ)	35.1	24.8

* ± refers to the standard deviation of the value.

3. Experimental Results and Discussion

3.1. Statistic Performance

The static performance, including the transfer, blocking, and output characteristics, is measured using a Keysight 1506A power semiconductor analyzer at room temperature. The results of one MOSFET from each cell type are shown in Figure 2. The threshold voltages (V_{th}) of all packaged MOSFETs are extracted using the linear extrapolation method from the transfer curves, as shown in Figure 2a. The average value and the standard derivation are shown in Table 1. A minimal V_{th} difference (<0.3 V) is observed between the MOSFETs with Dod and Oct cells. The transconductances (g_m) are also plotted in Figure 2a. The peak g_m value for the Dod-cell MOSFET is 1.5× higher than the fabricated Oct-cell MOSFET, indicating a higher current-driving capability. The blocking characteristics (Figure 2b) show less than 1 nA leakage current up to 600 V for both the Dod-cell and Oct-cell MOSFETs. The breakdown voltage (BV) is defined at the drain current of 100 μA. Similar average BVs and BV standard derivations are obtained for the Oct-cell and Dod-cell MOSFETs, as shown in Table 1. Therefore, the Dod cell does not degrade the blocking capability of the fabricated power MOSFET.

The output characteristics are shown in Figure 2c. A 2× higher drain current is observed on the fabricated Dod-cell MOSFET than on the Oct-cell MOSFET, which can be explained by the higher channel density and JFET density of the Dod cell. A higher channel density (total channel width in a unit cell/unit cell area) implies a larger total channel width and more current conduction within a specific die size. The channel densities for the Dod and Oct cells are calculated according to the geometry with the known cross section and cell pitch. The results are listed in Table 1. A 1.6× higher channel density is obtained for the Dod cell than the Oct cell. The higher channel density allows more current conduction for the fabricated Dod-cell MOSFET, resulting in a lower $R_{on,sp}$ than the fabricated Oct-cell MOSFET. The higher JFET density (JFET area/unit cell area) indicates a lower JFET region resistance, which contributes to the lower $R_{on,sp}$ [10]. The $R_{on,sp}$ of all the measured MOSFETs is calculated at a drain–source voltage (V_D) of 1.5 V. The mean values and standard derivations are listed in Table 1. A 2.2× reduction in the $R_{on,sp}$ is achieved for the Dod-cell MOSFET compared to the fabricated Oct-cell MOSFET.

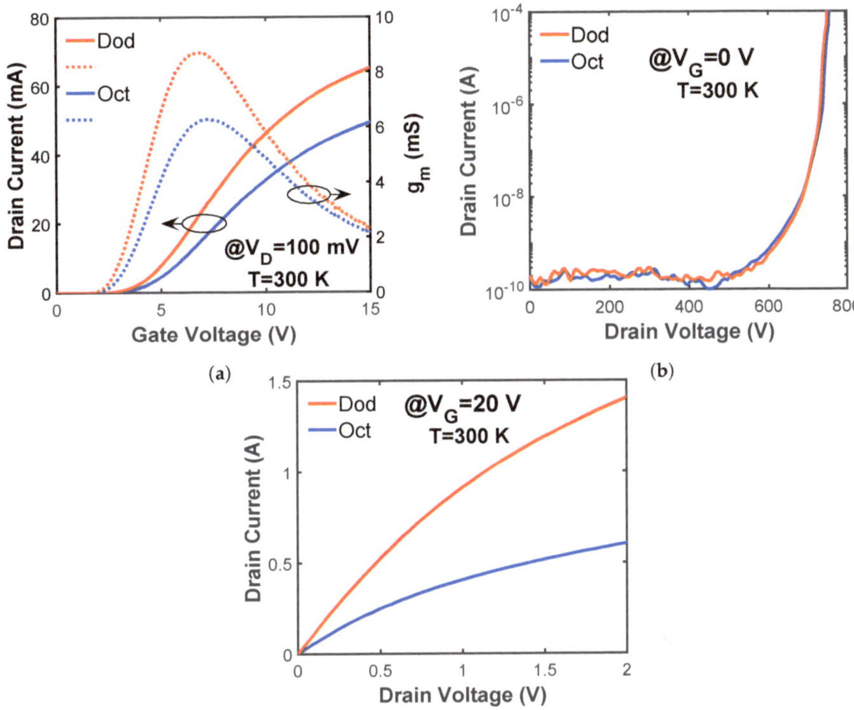

(c)

Figure 2. (**a**) Transfer characteristics at V_D = 100 mV (solid line: drain current vs. gate voltage; dash line: transconductance vs. gate voltage), (**b**) blocking characteristics at V_G = 0 V, and (**c**) output characteristics at V_G = 20 V, for the 650 V SiC power MOSFETs with Dod and Oct cells.

3.2. Dynamic Performance

The C_{gd} vs. drain voltage characteristics are shown in Figure 3. The C_{gd} significantly affects the dynamic behavior of SiC MOSFETs due to the well-known Miller effect [11,12]. The C_{gd} of a SiC planar MOSFET consists of the gate oxide capacitance on the top of the JFET region and the depletion capacitance of the JFET region and drift layer [13]. The results in Figure 3 show that the Dod-cell MOSFET has a higher C_{gd} than the fabricated Oct-cell MOSFET under any drain bias. Because the Dod-cell and Oct-cell MOSFETs are fabricated following the same process flow, their gate oxide thickness, JFET doping profile, and drift-layer doping profile are the same. Thus, the difference in the C_{gd} is caused by the different overlap areas of the gate and drain terminals (JFET region area) within a die. The calculated JFET density of the Dod cell is 1.6× higher than the Oct cell, which causes a higher C_{gd} of the Dod-cell MOSFET. The C_{gd} values at V_D = 400 V are extracted for all the measured MOSFETs. Additionally, the HF-FOMs, which are defined as the product of R_{on} and C_{gd}, are calculated. The HF-FOM of the Dod-cell MOSFET is 2.1× smaller than the fabricated Oct-cell MOSFET, indicating a better high-frequency performance.

A Double-Pulse Test (DPT), which provides straightforward data on the switching behavior of SiC MOSFETs [14], is conducted to investigate the switching performance of the fabricated MOSFETs. The results are shown in Figure 4. During the turn-on process (Figure 4a), the Dod-cell MOSFET shows a shorter transient time (higher turn-on speed) than the fabricated Oct-cell MOSFET. The turn-on and turn-off dv/dt, switching energy, and total switching energy of the measured MOSFETs are extracted and listed in Table 1. The Dod-cell MOSFET has a higher dv/dt and less switching energy than the measured

Oct-cell MOSFET during both the turn-on and turn-off processes. The Dod-cell MOSFET obtains a 29% reduction in total switching loss compared to the measured Oct-cell MOSFET.

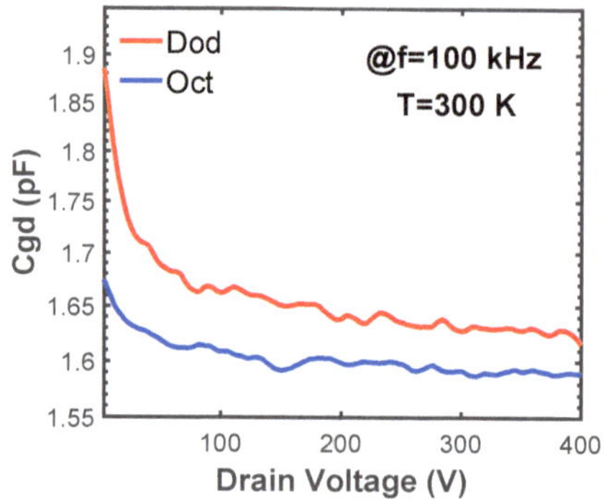

Figure 3. The C_{gd} of 650 V SiC MOSFETs with the Dod and the Oct cells.

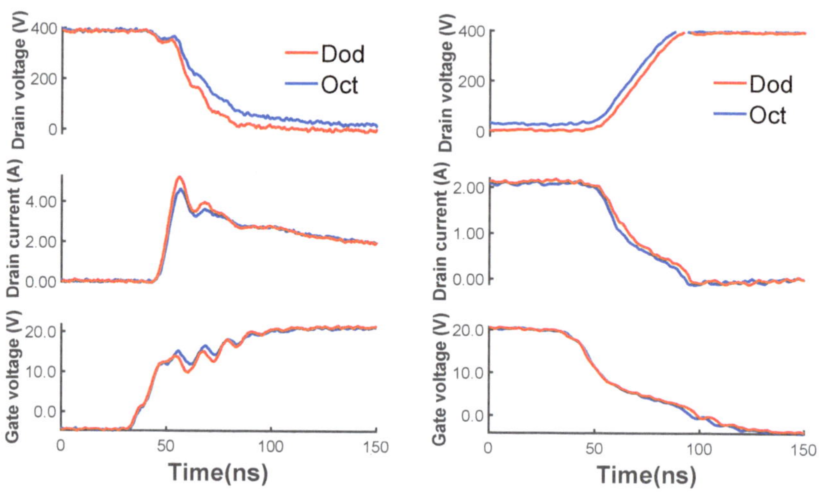

(**a**) (**b**)

Figure 4. (**a**) Turn-on and (**b**) turn-off waveforms of the fabricated 650 V SiC power MOSFETs with the Dod and the Oct cells.

4. Conclusions

A new cell topology named the Dod cell is proposed for SiC planar power MOSFETs. The Dod cell minimizes the JFET region area to achieve a low C_{gd} and improves the channel density to reduce the $R_{on,sp}$. A low C_{gd} indicates a higher switching speed and makes the Dod-cell MOSFET an attractive candidate for high-frequency applications. The reduced $R_{on,sp}$ decreases the conduction loss. The Dod cell and a recently published Oct cell (designed with minimal JFET region area) have been used for the layout design of 650 V SiC power MOSFETs. The fabricated Dod-cell MOSFET achieves a higher C_{gd}, but a 2.2× lower $R_{on,sp}$, 2.1× smaller HF-FOM, and 29% less switching losses compared with the fabricated Oct-cell MOSFET.

Author Contributions: Conceptualization, S.Z., T.L., and A.K.A.; methodology, S.Z., T.L., and A.K.A; investigation, S.Z. and J.F.; resources, A.S. and D.S.; data curation, S.Z.; writing—original draft preparation, S.Z.; writing—review and editing, T.L., M.H.W., and A.K.A. All authors have read and agreed to the published version of the manuscript.

Funding: This research is supported by the Block Gift Grant from the II-VI Foundation.

Institutional Review Board Statement: Not applicable.

Informed Consent Statement: Not applicable.

Data Availability Statement: Not applicable.

Acknowledgments: The authors would like to thank the Power Electronics Teams at Ford Motor Company's Research and Innovation Centers for helpful discussions.

Conflicts of Interest: The authors declare no conflict of interest.

Abbreviations

The following abbreviations are used in this manuscript:

SiC	Silicon Carbide
MOSFET	Metal-Oxide-Semiconductor Field-Effect Transistor
JFET	Junction Field-Effect Transistor
JBSFET	Junction Barrier Schottky (JBS) diode-integrated MOSFET
HF-FOM	High-Frequency Figure of Merit
DPT	Double-Pulse Test

References

1. Camacho, A.P.; Sala, V.; Ghorbani, H.; Martinez, J.L.R. A novel active gate driver for improving SiC MOSFET switching trajectory. *IEEE Trans. Ind. Electron.* **2017**, *64*, 9032–9042. [CrossRef]
2. Hazra, S.; De, A.; Cheng, L.; Palmour, J.; Schupbach, M.; Hull, B.A.; Allen, S.; Bhattacharya, S. High switching performance of 1700-V, 50-A SiC power MOSFET over Si IGBT/BiMOSFET for advanced power conversion applications. *IEEE Trans. Power Electron.* **2015**, *31*, 4742–4754.
3. Kimoto, T. Material science and device physics in SiC technology for high-voltage power devices. *Jpn. J. Appl. Phys.* **2015**, *54*, 040103. [CrossRef]
4. Baliga, B.J. Power semiconductor device figure of merit for high-frequency applications. *IEEE Electron Device Lett.* **1989**, *10*, 455–457. [CrossRef]
5. Wang, H.; Wang, F.; Zhang, J. Power semiconductor device figure of merit for high-power-density converter design applications. *IEEE Trans. Electron Devices* **2007**, *55*, 466–470. [CrossRef]
6. Watanabe, N.; Okino, H.; Shima, A. Impact of Cell Layout on On-state and Dynamic Characteristics of N-channel SiC IGBTs. In Proceedings of the 2022 IEEE 34th International Symposium on Power Semiconductor Devices and ICs (ISPSD), Vancouver, BC, Canada, 22–25 May 2022; pp. 85–88.
7. Agarwal, A.; Han, K.; Baliga, B.J. Impact of cell topology on characteristics of 600V 4H-SiC planar MOSFETs. *IEEE Electron Device Lett.* **2019**, *40*, 773–776. [CrossRef]
8. Agarwal, A.; Han, K.; Baliga, B. Assessment of linear, hexagonal, and octagonal cell topologies for 650 V 4H-SiC inversion-channel planar-gate power JBSFETs fabricated with 27 nm gate oxide thickness. *IEEE J. Electron Devices Soc.* **2020**, *9*, 79–88. [CrossRef]
9. Han, K.; Baliga, B. The 1.2-kV 4H-SiC OCTFET: A new cell topology with improved high-frequency figures-of-merit. *IEEE Electron Device Lett.* **2018**, *40*, 299–302. [CrossRef]
10. Liu, T.; Zhu, S.; Salemi, A.; Sheridan, D.; White, M.H.; Agarwal, A.K. JFET Region Design Trade-Offs of 650 V 4H-SiC Planar Power MOSFETs. *Solid State Electron. Lett.* **2021**, *3*, 53–58. [CrossRef]
11. Mukunoki, Y.; Nakamura, Y.; Konno, K.; Horiguchi, T.; Nakayama, Y.; Nishizawa, A.; Kuzumoto, M.; Akagi, H. Modeling of a silicon-carbide MOSFET with focus on internal stray capacitances and inductances, and its verification. *IEEE Trans. Ind. Appl.* **2018**, *54*, 2588–2597. [CrossRef]
12. Ebihara, Y.; Ichimura, A.; Mitani, S.; Noborio, M.; Takeuchi, Y.; Mizuno, S.; Yamamoto, T.; Tsuruta, K. Deep-P encapsulated 4H-SiC trench MOSFETs with ultra low R on Q gd. In Proceedings of the 2018 IEEE 30th International Symposium on Power Semiconductor Devices and ICs (ISPSD), Chicago, IL, USA, 13–17 May 2018; pp. 44–47.
13. Zhu, S.; Liu, T.; Fan, J.; Maddi, H.L.R.; White, M.H.; Agarwal, A.K. Effects of JFET Region Design and Gate Oxide Thickness on the Static and Dynamic Performance of 650 V SiC Planar Power MOSFETs. *Materials* **2022**, *15*, 5995. [CrossRef] [PubMed]
14. Li, X.; Jiang, J.; Huang, A.Q.; Guo, S.; Deng, X.; Zhang, B.; She, X. A SiC power MOSFET loss model suitable for high-frequency applications. *IEEE Trans. Ind. Electron.* **2017**, *64*, 8268–8276. [CrossRef]

materials

MDPI

Review

Novel Photonic Applications of Silicon Carbide

Haiyan Ou [1,*], Xiaodong Shi [1], Yaoqin Lu [1], Manuel Kollmuss [2], Johannes Steiner [2], Vincent Tabouret [3], Mikael Syväjärvi [4,*], Peter Wellmann [2,*] and Didier Chaussende [3,*]

[1] Department of Electrical and Photonics Engineering, Technical University of Denmark, Ørsteds Plads, Building 343, 2800 Kongens Lyngby, Denmark
[2] Crystal Growth Lab, Materials Department 6 (I-Meet), FAU Friedrich-Alexander University Erlangen-Nürnberg, Martensstr. 7, D-91058 Erlangen, Germany
[3] Université Grenoble Alpes, CNRS, Grenoble INP, SIMaP, 38000 Grenoble, France
[4] Alminica AB, Åsorp 2, 59053 Ulrika, Sweden
* Correspondence: haou@dtu.dk (H.O.); mikael@alminica.se (M.S.); peter.wellmann@fau.de (P.W.); didier.chaussende@grenoble-inp.fr (D.C.)

Abstract: Silicon carbide (SiC) is emerging rapidly in novel photonic applications thanks to its unique photonic properties facilitated by the advances of nanotechnologies such as nanofabrication and nanofilm transfer. This review paper will start with the introduction of exceptional optical properties of silicon carbide. Then, a key structure, i.e., silicon carbide on insulator stack (SiCOI), is discussed which lays solid fundament for tight light confinement and strong light-SiC interaction in high quality factor and low volume optical cavities. As examples, microring resonator, microdisk and photonic crystal cavities are summarized in terms of quality (Q) factor, volume and polytypes. A main challenge for SiC photonic application is complementary metal-oxide-semiconductor (CMOS) compatibility and low-loss material growth. The state-of-the-art SiC with different polytypes and growth methods are reviewed and a roadmap for the loss reduction is predicted for photonic applications. Combining the fact that SiC possesses many different color centers with the SiCOI platform, SiC is also deemed to be a very competitive platform for future quantum photonic integrated circuit applications. Its perspectives and potential impacts are included at the end of this review paper.

Keywords: silicon carbide; integrated photonics; material growth

Citation: Ou, H.; Shi, X.; Lu, Y.; Kollmuss, M.; Steiner, J.; Tabouret, V.; Syväjärvi, M.; Wellmann, P.; Chaussende, D. Novel Photonic Applications of Silicon Carbide. *Materials* **2023**, *16*, 1014. https://doi.org/10.3390/ma16031014

Academic Editors: Marilena Vivona and Mike Jennings

Received: 19 December 2022
Revised: 5 January 2023
Accepted: 17 January 2023
Published: 22 January 2023

1. Introduction

Photonic integrated circuits (PIC), also named integrated lightwave circuits (ILC) or photonic lightwave circuits (PLC), have been discussed to overcome the predicted bottleneck for integrated circuits for more than half a century. During the development of PIC, a lot of different material platforms have been investigated, such as silicon (Si), silicon dioxide (SiO_2), silicon nitride (SiN), aluminum gallium arsenide (AlGaAs), indium phosphide (InP), and the emerging gallium nitride (GaN), lithium niobate ($LiNbO_3$), silicon carbide (SiC), etc. Although Si is playing the most important role in this area, benefiting from the well-established complementary metal-oxide-semiconductor (CMOS) processing, it is limited by its own optical properties, such as indirect bandgap, strong two-photon absorption at telecommunication wavelength and zero second-order nonlinearity. Therefore, other material platforms are developed either monolithically on their own or as hybrid platforms that are integrated with an Si substrate. Among them, SiC is emerging rapidly in this field, partially because of the commercially available high-quality crystal SiC wafers and mature micro and nanofabrication of SiC in the promotion of SiC powered electronic devices, a similar story as Si. Unlike many well defined figures of merit (FOM), such as Johnson FOM, Keyes FOM, Baliga FOM, and Baliga high-frequency FOM, for direct comparison of different material platforms in electronics, photonic material platforms are usually compared in terms of bandgap, refractive index, second-order nonlinearity, and third-order nonlinearity.

Normally, for an ideal PIC material platform, it has low loss at the working wavelength; has a high refractive index for strong light–matter interaction; has strong second-order nonlinearity for electro-optical modulation and second-harmonic generation; and has high third-order nonlinearity for efficient wavelength conversion. As shown in Table 1, SiC fulfills all the demanded optical properties: high refractive index, wide bandgap, and high second-order and third order nonlinearities. Additionally, SiC is CMOS compatible. Therefore, extensive investigations on SiC optical devices are emerging.

Table 1. Main material platforms for PICs. n: refractive index, BG: bandgap, n_2: nonlinear refractive index, $\chi^{(2)}$: second-order susceptibility.

Materials	Key Features				Pros	Cons
	n	BG (eV)	n_2 (m^2/W)	$\chi^{(2)}$ (pm/V)		
Si [1–5]	3.48	1.12	4.5×10^{-18}		High refractive index, compact, CMOS compatible, cost-effective	Small bandgap, strong two photon absorption
SiO_2 [6,7]	1.5	9	3×10^{-20}		CMOS incompatible, low loss	Low refractive index, low nonlinearity
SiN [8–10]	2.0	5	2.5×10^{-19}		CMOS compatible, cost-effective	Low nonlinearity
Hydex [11]	1.7	1.8	1.2×10^{-19}		CMOS compatible, cost-effective	Low nonlinearity
GaP [12–14]	3.1	2.26	1.2×10^{-17}	200	High refractive index, high nonlinearity	Cost-ineffective, CMOS incompatible
AlN [15,16]	2.1	6	2.3×10^{-19}	1	Efficient light source	Cost-ineffective, CMOS incompatible
Diamond [17]	2.4	5.5	8.2×10^{-20}		CMOS compatible, wide bandgap	Cost-ineffective, difficult to etch
AlGaAs [18,19]	3.3	1.4–2	2.6×10^{-17}	100	High refractive index, high nonlinearity	Small bandgap, cost-ineffective, CMOS incompatible
$LiNbO_3$ [20–22]	2.2	4		27	Well-known optical properties, cost-effective	CMOS incompatible, difficult to etch
SiC [23]	2.6	2.4–3.2	1.3×10^{-18}	33	High refractive index, compact, CMOS compatible, high nonlinearity, light emitter	Cost-ineffective

2. Silicon Carbide Photonic Integrated Platforms

Planar optical waveguides are key structures where the light is confined and transmitted by obeying total internal reflection (TIR). For SiC waveguides, it is indispensable to form an index contrast where SiC has a higher refractive index than the surroundings. At the beginning of this research, floating SiC in the air has been applied because the commercial SiC epilayers on Si substrates were available. Since Si has a higher refractive index than SiC, it has to be removed near the SiC waveguides. But this structure has restrictions on the post-etching process, such as patterning and top-cladding deposition. Instead, silicon carbide on insulator stacks is developed and applied in most of the publications currently. There are generally three different methods to form SiCOI, as shown in Figure 1. For method 1, smart/ion cut method is adopted from silicon-on-insulator (SOI) stack's formation, where Si wafer is replaced by a SiC wafer [24,25]. But since SiC and Si have very different physical properties, the material loss in SiCOI is still quite high compared to SOI, after ion implantation. This method mainly delivers 4H SiCOI stacks. For method 2, the bonded SiC is thinned down directly by using grinding and later chemical mechanical polishing (CMP) to smoothen the surface. In this way, the ion implantation step in the smart/ion cut is skipped and SiCOI stacks achieve very low loss. This method delivers 4H and 3C SiCOI stacks. For method 3, amorphous SiC is directly deposited on the Si substrate with an SiO_2 layer between. The amorphous SiC deposition could be completed through plasma-enhanced chemical vapor deposition (PECVD) and sputtering. Both methods are CMOS compatible.

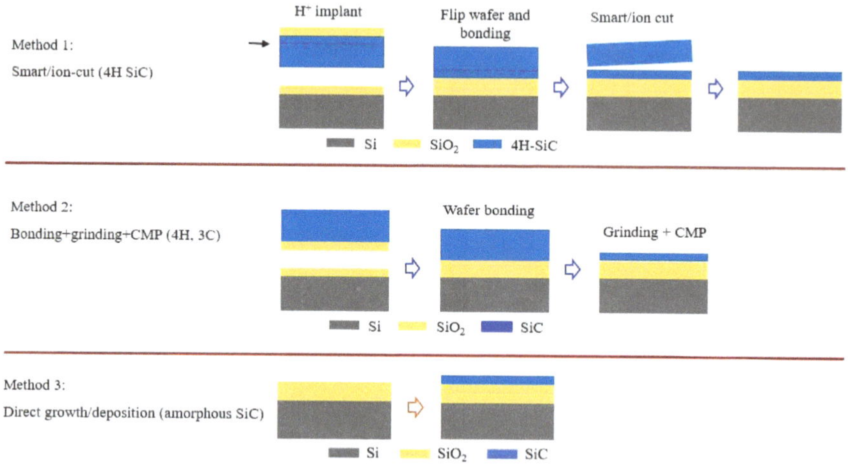

Figure 1. Three SiCOI stack formation methods.

Table 2 listed the state of the art of SiCOI waveguides in terms of waveguide structure, dimension, forming method, propagation loss, and quality factor (Q) of a microring resonator and the measured nonlinear refractive index. The active SiC layers have a thickness of hundreds of nm, which is extremely difficult to control in the process of thinning down SiC from hundreds of µm to nm. So far, 4H SiCOI waveguides formed from method 2 show the lowest loss of 0.38 dB/cm. 3C SiCOI from method 2 and amorphous SiC from method 3 have quite close losses of about 3 dB/cm. The loss of 4H SiCOI formed by method 1 could be derived from the quality factor of the microring, which is about 6.5 dB/cm. The waveguide loss is related to both the material loss and the waveguide fabrication imperfection. Nanofabrication technology has been well developed for other material platforms, such as SOI. SiC could be etched by using the same machine and precursors as Si. So, the fabrication-related loss is reduced to a negligible level for SiC in a very short time, while the supply of low-loss SiCOI stacks remains challenging. Comparing the listed 3 methods, method 2 shows the best performance, while method 1 and 3 have the similar performance, currently. However, with regard to the cost, method 2 is the most expensive one, followed by method 1 and 3 with reduced cost. In addition to the unique optical properties, SiC has the potential for easy scale-up. From this point of view, method 3 shows advantages because it is not limited by the rather small SiC wafers available, unlike method 1 and 2. Looking forward to quantum applications of SiCOI, many optically addressable defects in crystalline SiC such as 4H, 3C and 6H have been published [26,27], which are promising as candidates for single photon sources. But no report on optically addressable defects from amorphous SiC have been revealed yet. 6H SiCOI has been reported for usage in optical cavities [28–32]. It is used less, as 4H SiC is becoming dominant because of its better crystal quality, which is crucial for high quality-factor optical cavities. Moreover, many physical and optical properties of 6H SiC lie between 3C and 4H SiC, making it less interesting.

Table 2. State of the art of SiC waveguides.

Material	Stack	Method	Size (nm^2)	Loss (dB/cm)	Q	n_2 (m^2/W)
3C SiC [33]	SiC-Si	Floating	730 × 500			5.3×10^{-19}
3C SiC [34]	SiCOI	2	1700 × 500	2.9	1.42×10^5	
a SiC [35]	SiCOI	3	800 × 350	3	1.6×10^5	4.8×10^{-18}
4H SiC [36]	SiCOI	1	750 × 500		7.3×10^4	
4H SiC [37]	SiCOI	1	2700 × 600			8.6×10^{-19}
4H SiC [38]	SiCOI	2	3000 × 530	0.38	1.1×10^6	6.9×10^{-19}

3. Silicon Carbide Photonic Cavities

The light-matter interaction could be enhanced by an optical cavity. Normally, a higher quality factor and a smaller mode volume are preferred. As an example for a quantitative relation, power threshold for parametric oscillation is determined by the waveguide loss (described by the loaded quality factors $Q_{L,s}$, $Q_{L,i}$, and $Q_{L,p}$ of the signal, idler, and pump modes, respectively) and the confinement, given by

$$P_{th} = \frac{\omega_0 n^2}{8\eta n_2 c} \frac{V}{Q_{L,p}\sqrt{Q_{L,s}Q_{L,i}}} \quad (1)$$

where n is the modal refractive index, V is the mode volume, and $\eta = \frac{Q_{L,p}}{Q_{c,p}}$, where $Q_{c,p}$ accounts for coupling from the pump mode to the waveguide, n_2 is the nonlinear refractive index and ω_0 is the angular frequency [38]. For SiC, three types of optical cavities have been published: photonic crystal cavities (PhC), microring resonators (MRR) and microdisk resonators (MDR). Their evolvement during the past ten years is shown in Figure 2.

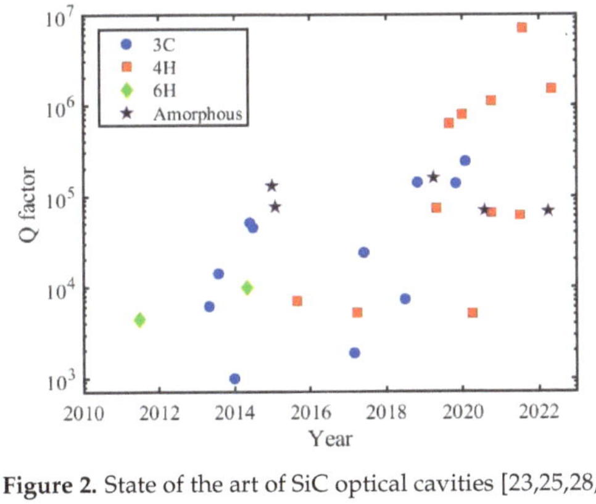

Figure 2. State of the art of SiC optical cavities [23,25,28,32,34–36,38–58].

Among the different forms of SiC (3C, 4H and 6H polytypes, and amorphous), there has been no further progress for 6H SiC in recent years. 4H SiCOI microring resonators made by method 2 have demonstrated a quality factor of more than 1 million. Based on it, advanced applications such as optical frequency combs have been demonstrated as well.

3.1. Microring and Microdisk Resonators

Thanks to the advanced nanofabrication technologies, SiC microresonators, including MRR and MDR, with low loss and high quality factors, have been demonstrated. The loss is jointly contributed by the material absorption loss and the light scattering loss due to the surface roughness. Figure 3 shows scanning electron microscope (SEM) images of several high-performance SiC microresonators.

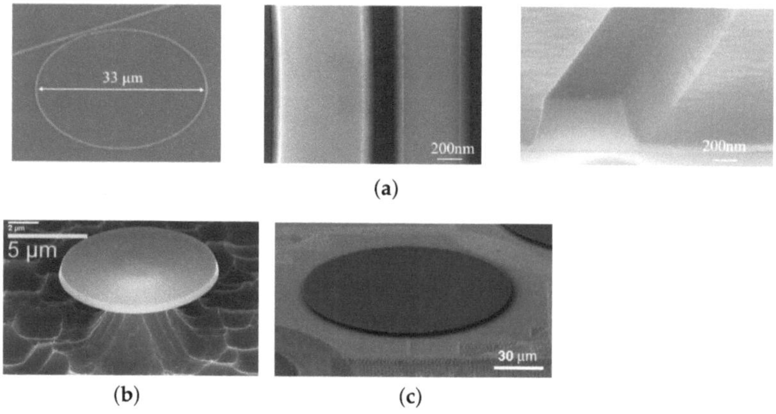

Figure 3. Scanning electron microscope images of (**a**) 4H SiCOI MMR, reprinted with permission from [36] ©The Optical Society, (**b**) 3C SiC MDR [59], and (**c**) 4H SiCOI MDR [56].

For 3C SiC, using the freestanding SiC, grown on the Si substrate, quality factors of 2.4×10^4 and 5.12×10^4 have been reported for MRR and MDR, respectively [42,49]. The high loss mainly comes from the material absorption and is associated to the high density of extended defects, inherently linked to the 3C-SiC/Si heteroepitaxial system, and discussed in the coming section. The highly defective layer, which is close to the 3C-SiC/Si interface, can be removed, if preparing 3C SiCOI stacks. As a result, microresonators with much higher quality factors can be achieved, which are 1.42×10^5 and 2.42×10^5, for MRR and MDR, respectively [34,53]. For 4H SiC thin film prepared through the smart/ion cut method, quality factors of 7.3×10^4 and 5.25×10^3 have been reported for MRR and MDR, respectively [36,60]. Chemical-mechanical polishing and post-oxidation processes can also help to reduce the roughness of the top surface and the sidewall of the waveguides, and thus improve the quality factors of SiC microresonators, for both 3C and 4H SiC [34,36,61,62]. For 4H SiC thin film prepared through the grinding and CMP method, ultra-high quality factors of 1.1×10^6 and 6.75×10^6 have been reported for MRR and MDR, respectively [38,56]. Regarding the amorphous SiC deposited by PECVD, the MRR with a quality factor as high as 1.6×10^5 has been reported [35]. Based on these demonstrations of high-performance SiC microresonators, as well as other SiC integrated devices, SiC integrated platforms can be potentially applied to classical and nonclassical optical communication systems for signal processing, logic gate, modulation, filtering, (de)multiplexing, etc. [5,24,63–68].

The field enhancement in the microresonators can be reflected by the photon density. With SiC microresonators, efficient nonlinear effects-based frequency conversion has been observed and studied, according to the material-based second- and third-order nonlinearities in SiC. Second harmonic generation in 4H SiCOI MRRs has been demonstrated with sub-milliwatt power, through the mode-phase-matching method [52]. Besides, basic third-order nonlinear phenomena, such as self-phase modulation, four-wave mixing, cascaded four-wave mixing, and optical parametric oscillation have also been demonstrated in 3C, 4H, and amorphous SiC or SiCOI microresonators [33,35,37,38,55,62,69,70]. Through these experiments, the nonlinear refractive index of 3C and amorphous SiC has been extracted to be 5.3×10^{-19} m^2/W, and 4.8×10^{-18} m^2/W, respectively [33,55]. The birefringence of the third-order nonlinearity of 4H SiC has been experimentally demonstrated, the nonlinear refractive index along the extraordinary and the ordinary axis of 4H SiC is measured to be 1.31×10^{-18} m^2/W and 7.0×10^{-19} m^2/W, respectively, which indicates the transverse magnetic polarization can perform more efficient wavelength conversion [23]. Moreover, classical and quantum frequency combs in 4H SiCOI MMR and MDR with ultra-high quality factors are observed, including octave-spanning Kerr frequency combs, Raman combs, and quantum solitons, which are shown in Figure 4 [56,57,71]. The excellent optical nonlinear properties of SiC, as well as these nonlinear phenomenon demonstrations, make SiC integrated platforms possible for plenty of nonlinear optical applications, such as light sources with multiple wavelength for optical communication, photon-pair sources for quantum computing and communication, spectroscopy, metrology, sensing, etc. [3,4,72].

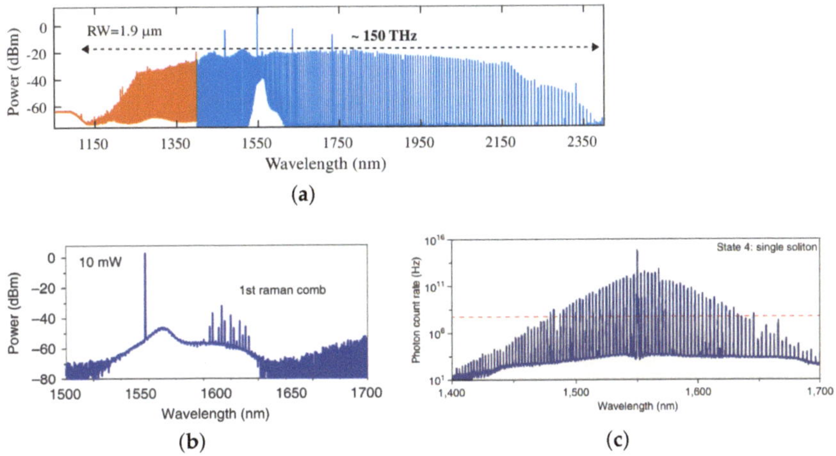

Figure 4. Experimental spectra of (**a**) octave-spanning Kerr frequency comb reprinted with permission from [57], (**b**) Raman comb, reprinted with permission from [56], and (**c**) quantum soliton comb, reprinted with permission from [71] in SiC microresonators.

Additionally, the thermo-optic behaviors of the MRRs in 3C, 4H and amorphous SiC integrated platforms have been studied. The external temperature induced resonance shift and the material absorption-induced bistability in 4H and amorphous SiCOI are investigated [55,58,62]. Micro-heaters have been integrated with the 3C SiCOI MRR for efficient thermo-optic phase shifting [51]. Meanwhile, the thermo-optic coefficients of 3C, 4H and amorphous SiC are measured to be 2.67×10^{-5}/K, 4.21×10^{-5}/K, and 1.4×10^{-4}/K, respectively [51,55,62]. Devices made of 3C and 4H SiC that have small thermo-optic coefficients, can exhibit high thermal stability, but amorphous SiC, having a large thermo-optic coefficient, allows efficient modulation. The electro-optic effect in 3C SiCOI has also been observed, and the Pockels coefficient is extracted to be 2.6 pm/V, which enables high-speed electro-optic modulation [73,74]. An all-optical Kerr switching intensity modulator at 25 Gbps in amorphous SiCOI has been reported [75].

3.2. Photonic Crystal Cavities

Besides the microring and microdisk, the cavities realized by photonic crystal structures are also widely studied in the SiCOI platform. Because of the potential of emitting single photons by defects in SiC, photonic crystal cavities are studied to enhance the emission rate of the defects by addressing the defects in the cavities. The amplified magnitude of the modified emission rate comparing with the defects in free space is expressed by the Purcell factor in Equation (2) [76]. In this equation, F_p is the Purcell factor, Q is the quality factor of a certain cavity mode, V_0 is the mode volume, and λ is the wavelength of light transmitting in the dielectric material.

$$F_p = \frac{3Q\lambda^3}{4\pi^2 V_0} \quad (2)$$

In order to realize a large increase of the emission rate of photons from defects in SiC, the strong light–matter interaction is required. From Equation (2), the quality factor of the photonic crystal cavity should be as large as possible, and the mode volume is preferred to be kept small. Due to the large scattering loss, the quality factor of the PhC cavity is much smaller than that in the microring or microdisk.

The photonic crystal can be briefly classified into one-dimensional (1D) photonic crystal and two-dimensional (2D) photonic crystal in SiCOI platform. A 1D photonic crystal waveguide is also called a photonic crystal nanobeam. As long as the holes are modified in terms of positions or radius, a nanocavity will form and can support several modes in the nanobeam. A 2D photonic crystal can further be transformed into a 2D photonic crystal cavity by removing several holes in the center, such as one hole (H1 cavity) or three holes (L3 cavity), or a 2D photonic crystal waveguide by removing a line of holes. By tuning the lattice constant along the photonic crystal waveguide direction in several periods, a heterostructure cavity will be formed [77]. Regarding the application of color center photoluminescence enhancement, a 1D nanobeam is preferred to 2D photonic crystal structures because of its lower modal volume and the feasibility of the undercut to have a larger refractive index contrast.

For 6H SiC prepared by smart-cut, the L3 cavity and heterostructure cavity were firstly fabricated in a smart-cut SiCOI platform [28]. The quality factor of an optimized L3 cavity of 2×10^3 and the quality factor of the heterostructure cavity of 4.5×10^3 have been reported. These two kinds of cavities on 6H SiCOI were improved to 1×10^4 quickly afterwards [29–31]. A second harmonic generation has been realized in a photonic crystal heterostructure cavity in the same platform [32]. For 3C SiC, the L3 cavities were realized with an initial quality factor of around 0.8×10^3 in an epitaxially grown thin film [41]. The more detailed results have been reported with the quality factor of 1.5×10^3 for L3 and around 1×10^3 for H1, despite a theoretical prediction of 1.7×10^4 and 4.5×10^4, respectively [43]. The imperfection of fabrication and deviation of the hole radius or positions are likely to cause a low quality factor. For amorphous SiC, the highest reported experimental quality factor of 1D photonic crystal cavity is 7.69×10^4, and the mode volume is around $0.6 \times (\lambda/n)^3$ [45]. The Purcell factor is calculated to be around 10^4. This cavity is a suspended 1D photonic crystal nanobeam cavity. The decline of experimentally measured quality factor, compared with a simulation result of 1.75×10^6, is mainly from the scattering of imperfect sidewalls. For 4H SiC, the high experimental quality factor of 6.3×10^5 is achieved in a ground SiCOI platform in a heterostructure cavity [50]. A photonic crystal nanocavity with such a high quality factor enables strong second-harmonic generation conversion efficiency. The scheme of the structure and the SEM images are shown in Figure 5b. In a 1D photonic crystal nanobeam cavity, the quality factor is smaller: 1.93×10^4 [52]. The color centers inside the nanobeam cavity can enhance the emission efficiency; the structure is shown in Figure 5a. All these works point out the great reduction of experimental quality factor as compared with simulated or calculated quality factor. The non-vertical sidewall angle is the main cause of this difference. The other attributions include the sidewall roughness, deviation of hole radius or position, and material absorption.

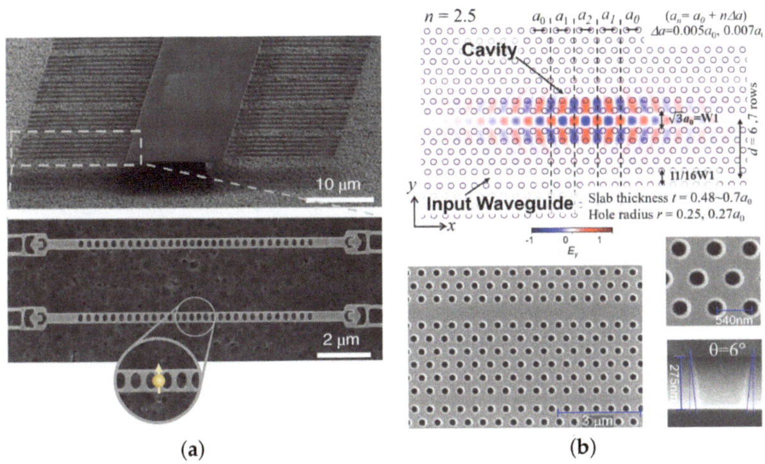

Figure 5. Scanning electron microscope images of (**a**) 1D photonic crystal nanobeam cavities, reprinted with permission from [52] and (**b**) 2D photonic crystal heterostructure cavity, reprinted with permission from [50] ©The Optical Society.

4. Silicon Carbide Materials

Different from the electronic devices, the material requirements for SiC photonic applications handle surface roughness, film thickness and uniformity in nanometer scale, which poses big challenges on the material quality, transfer and processing. As mentioned in Section 2, 4H, 3C and amorphous SiC have shown loss at different levels. If we assume the nanofabrication has been optimized and the fabrication-related loss is negligible, the loss level indeed reflects the material loss. The material loss consists of absorption and the scattering of defects in the material. The most promising large-area thin-film-deposition processes (amorphous and micro-crystalline material) are of particular interest due to their huge variability in terms of stacking thin optical films on almost any kind of substrate and because of their comparably low manufacturing costs. The currently mid-range optical loss may be reduced by further improvement of the deposition processes. Therefore, in the following an overview on all SiC-based thin film methods currently under development will be given. The lowest optical loss may be achieved in single-crystalline SiC material exhibiting a dislocation density that is as low as possible, as well as unintentional doping. At the same time, by adaption of the single-crystal growth process, intrinsic point defects may be generated for single-photon-source or qubit applications. Besides the preparation of high-quality 4H-SiC crystals and wafers, also bulk growth of 3C-SiC will be outlined. The latter material exhibits a higher defect density than 4H-SiC. However, its higher crystalline symmetry may be of advantage at a certain stage if optical active centers (point defects and intentional doping) are added as a functionality to the electrical-optical devices.

4.1. Amorphous Silicon Carbide

Contrary to the SiC polytypes, which have a well-defined crystalline structure and composition, amorphous Silicon Carbide (a-SiC) is a very versatile material. It can vary in stoichiometry, such as the Si:C ratio, in density, and of course in the resulting optical, electronic and mechanical properties. Depending on the deposition method, it can also contain hydrogen (H:SiC), which provides even further tuning of the properties. At the first order, the primary parameter which determines the formation of amorphous SiC is the temperature, as it enhances the relaxation of the "structure" towards crystalline materials. There is thus a temperature threshold above which crystallization will systematically occur. This temperature is not precisely determined in the literature as it also depends on secondary parameters like supersaturation, reaction pathway, presence of impurities and so on. It is nevertheless strictly inferior to 800 °C. We review in the following the low temperature, vapour-phase deposition methods that allow the formation of amorphous SiC thin films.

4.1.1. Low Temperature Chemical Vapour Deposition

A first process which can lead to amorphous SiC is the Low Temperature Chemical Vapour Deposition. Using standard chemistry (Si-C-H or Si-C-H-Cl), such as is discussed in the next section for epitaxial growth, it is not possible to deposit in the targeted temperature range, as the gaseous species are too stable. Differently said, the reaction kinetics are infinitely slow. However, with organometallic precursors such as monomethylsilane (MMS), a deposition temperature lower than 650 °C and down to room temperature can form a-SiC [78–80].

Contrary to the standard CVD, where the initial cracking of the precursors in the gas phase provides reactive intermediates, the MMS is directly adsorbed on the surface of the sample. For that, the surface reactivity is promoted either by the deposition of a hydrogenated silicon layer or by a plasma etching [81], which enhances the adsorption efficiency. The hydrogen atoms get desorbed during this reaction. However, because of the low temperature of the process and the needed hydrogen desorption, the stoichiometry of the SiC becomes hard to control [80,81]. No optical study using films deposited by this process was found in the literature.

4.1.2. Plasma-Enhanced Chemical Vapour Deposition

Plasma-Enhanced CVD (PECVD) is probably the most widely used technique for the deposition of a-SiC thin films [82,83]. The substrate is kept at a temperature lower than 500 °C, and often even lower than 300 °C. The chemical reactions are induced in a plasma (usually in Argon and/or hydrogen), which drastically increases the reactivity of the intermediate gaseous species. The ionic species produced are then adsorbed on the surface, whereas the organic wastes are evacuated. The energy of the ion bombardment allows the migration of the adsorbed species and the formation of dense layers. The use of hydrogen as carrier gas systematically leads to the hydrogenation of the SiC films. This limits the maximum refractive index achievable, which is typically around 2.36 [83]. The refractive index decreases with the density, which can be tuned by varying the plasma power [83–86], as shown in Figure 6.

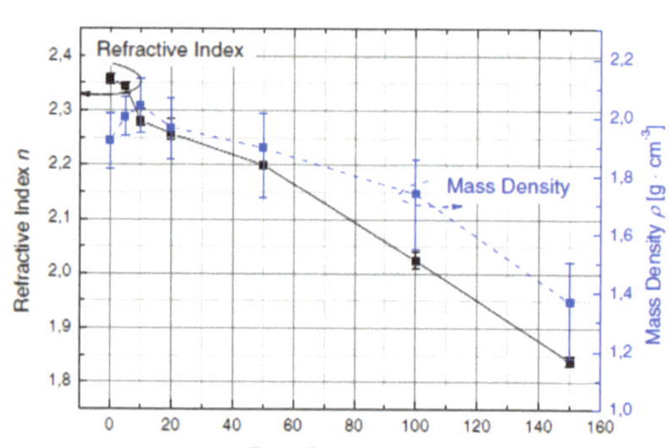

Figure 6. Evolution of the refractive index (left) and the density (right) as a function of the power applied on the substrate P_T measured by Frischmuth team [83]. Reprinted with permission from Elsevier.

An increase of the substrate temperature and of the pressure cause an increase of the refractive index [85,86] and the decrease of the roughness down to 1 nm [86].

4.1.3. Hot Wire Chemical Vapour Deposition

Known as one of the pioneer CVD methods, Hot Wire CVD (HWCVD), also called Hot Filament CVD (HFCVD) is also well adapted for the deposition of amorphous SiC [87–89]. In this case, the energy necessary to crack the precursors and generate highly reactive intermediates is provided by a metallic filament composed of Rhenium, Tantalum or Tungsten placed upstream in the reaction chamber and heated at high temperature, typically around 2000 °C [90,91]. The thermally activated species are then adsorbed on the substrate, which can be kept at a low temperature, even lower than 500 °C. The presence of hydrogen radicals (H*) generated at high temperature by the filament produces an etching reaction of the surface, which can remove the adsorbed species located in unstable positions. A high concentration of H* can thus produce crystalline layers, with the possibility to tune the crystallite size with the H* concentration [87,92–94]. The formation of a stoichiometric, crystalline SiC layer is ensured by decomposing the carbon-containing precursor (typically CH_4) at a temperature higher than 2000 °C [93]. At a temperature lower than 2000 °C, with low pressure and a low amount of hydrogen, the formation of an amorphous SiC layer can be obtained. Similarly to PECVD, this technique causes the incorporation of hydrogen inside the amorphous SiC film, decreasing the refractive index. The roughness of the SiC deposit can be also very low, in the nm range [95].

4.1.4. Radio Frequency Magnetron Sputtering

Amorphous SiC can be deposited with a magnetron sputtering method. In this process, one or several targets with the composition aimed are eroded with a plasma. Then, the ionized species formed are adsorbed on the surface of the sample [96–100]. Non-conductive compounds used as targets, such as SiC or Si, require the use of Radio Frequency Magnetron Sputtering to prevent the accumulation of electric charge [101]. Amorphous layers can be deposited with a temperature lower than 500 °C [102,103]. As a rule, a low deposition temperature increases the refractive index up to 3.2 at a wavelentgh of 630 nm [103,104]; whereas it diminishes the transmittance of amorphous SiC [104], as shown in Figure 7.

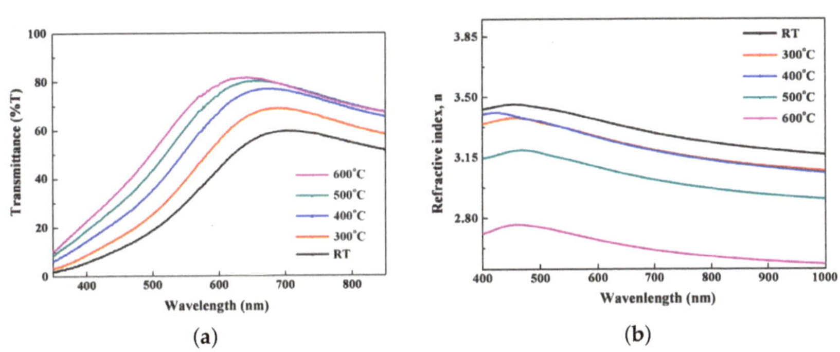

Figure 7. Evolution of (**a**) the transmittance and (**b**) the refractive index as a function of the temperature of the substrate, after Seo team [104]. Reprinted with permission from Elsevier.

Moreover, lowering the electrical power applied on the target increases the transmittance and diminishes the concentration of carbon clusters [103,105], as shown in Figure 8. Tang reported that decreasing the power diminishes the size of the particles and modifies their shapes from a dense globular-shaped structure to columns. This change of particle morphology diminishes the roughness [106]. The optical gap is evaluated at 3.5 eV [105,107], but the presence of graphite in the amorphous SiC can put the gap between 1.3 to 1.7 eV [103,105,108].

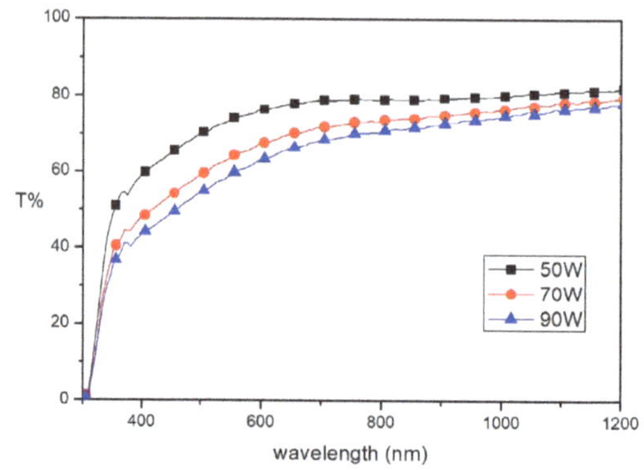

Figure 8. Evolution of the transmittance of SiC films in function of the RF power, measured by Wang [105]. Reprinted with permission from Elsevier.

4.2. *3C Silicon Carbide*

4.2.1. Growth of 3C-SiC-on-Si Heteroepitaxy

Compared to its hexagonal counterparts (4H- and 6H-SiC) for the cubic polytype (3C-SiC), bulk growth using Physical vapour transport (PVT) is not established as "standard" method yet, which is mainly attributed to the lack of suitable seeding substrates. However, 3C-SiC is the only polytype that can be nucleated using heteroepitaxial chemical vapour deposition (CVD) on Si. Moreover, due to the higher process control available in CVD, the fine tuning of material properties especially the doping concentration is increased compared to PVT systems, especially for the goal of low-doped material. While silicon as a starting material can lower the cost for device fabrication due to the availability of large area substrates, a series of challenges have to be solved. The two main obstacles to overcome are the difference in the lattice constant of approx. 20% and the thermal expansion coefficient (TEC) of ca. 8% at room temperature between 3C-SiC and Si [109,110]. These misfits are responsible for a variety of defects formed at the interface, e.g., stacking faults (SFs), anti-phase boundaries (APB) or even three-dimensional defects such as protrusions [111] and additionally result in wafer-bending after the cool-down to room temperature [112]. The two most common defects in heteroepitaxial growth of 3C-SiC on Si are SFs and APB. SF along the {111} planes are mostly described as an intrinsic defect, based on the lattice mismatch and difference in TEC between 3C-SiC and Si. However, a decrease in the SF density could be achieved by annihilation during growth of thicker 3C layers [113] or advanced growth methods, such as the "switch back" epitaxy on undulated substrates described by Nagasawa et al. [114]. Different APBs are 2D defects that are formed by two 3C-SiC crystals, which are rotated 180° around the [110] axes. While getting rid of SF completely is somehow still impossible, the application of an off-axis Si substrate can lead to an APB-free material, as described in [115]. Today, the "standard" heteroepitaxial CVD growth process consists of a hydrogen etching step to remove the native oxide layer from the Si substrate, the growth of a buffer layer or carbonization layer and the main growth step, using Si and carbon (C) precursors followed by the cool down of the system. Traditionally silane (SiH_4) and propane (C_3H_8) or ethylene (C_2H_4) are used while hydrogen (H_2) acts as the carrier gas. However, a variety of different approaches can be found in the literature using different precursors or adding additional gas species. Often chlorine is added either in the form of hydrogen chlorine (HCl) or chlorine-based precursors like trichlorosilane (HCl_3Si) to improve the material quality and increase the growth rate [116]. One of the most important aspects regardless of the used gases is the C/Si ratio during the main growth step. Tuning this parameter has not only an influence on the growth rate but also strongly influences the material quality. Chassagne et al. showed that for too-low C/Si ratios, the layer morphology and crystalline quality quickly degrade [117]. Meanwhile for too high ratios, homogenous gas phase reactions can occur leading to SiC clusters on the surface. Besides the correct adjustment of the C/Si ratio, the formation of the mentioned buffer layer is crucial when it comes to material quality. The first vital studies on this topic were conducted by Nishino et al. [118] and later by Liaw and Davis [112]. Both groups described the importance of a buffer layer formed by introducing a carbon-containing precursor prior to the main growth step during the heat up of the system. During this step the gaseous carbon species will produce an initial layer of SiC by reacting with the Si of the substrate. Due to the formation of this buffer layer the lattice mismatch between 3C-SiC and Si can be reduced significantly, leading to lower defect densities and increased material quality. A setback of the buffer layer is the formation of voids at the interface 3C-SiC/Si. Several theories can be found in the literature explaining this phenomenon. Most likely, the formation of voids is attributed to the surface diffusion of Si [119]. After the initial 3C-SiC seeds are formed due to the deposited carbon, Si is removed from the open substrate surface between these seeds to supply the formation of 3C-SiC, resulting in the observable voids (see Figure 9). Since voids can locally favor the formation of dislocations and SFs, a lot of research has been conducted to date to improve the buffer layer quality and reduce void formation. Possible approaches include the variation of the C/H_2 ratio [120,121], the process pressure [122] or the initiation of Si precursors flow during the formation of the buffer layer [123].

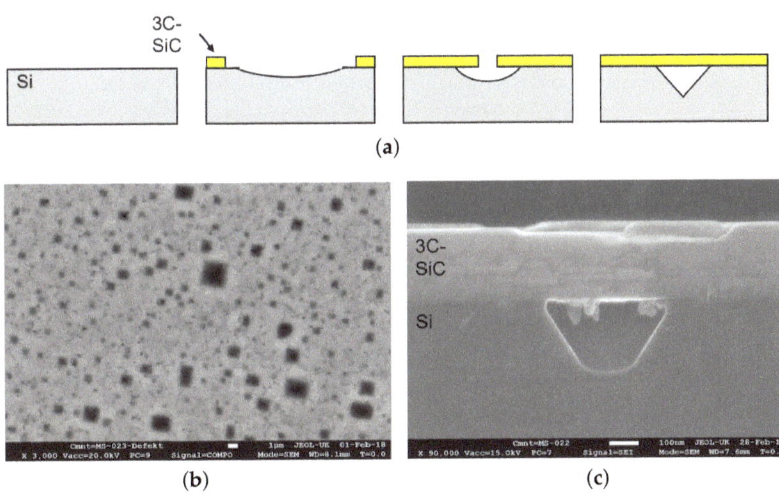

Figure 9. (**a**) Scheme of the formation of voids at the 3C-SiC/Si interface. (**b**) Top view scanning electron microscopy (SEM) image of a 3C on Si sample grown at 1200 °C. The darker areas are attributed to voids underneath the 3C-SiC layer. (**c**) Crosscut SEM image of a void at the Interface 3C-SiC/Si.

Based on the lattice mismatch between 3C-SiC and Si, additional focus has to be given to the management of thermal stress induced by the difference of TECs between Si and 3C-SiC, leading to wafer-bending after the cool down from growth temperature. Zielinski et al. [109] also showed that the TEC mismatch cannot be changed; the final wafer-bow of heteroepitaxial grown 3C on Si can be controlled based on the C/Si ratio, final layer thickness or the growth rate during the deposition. Another approach was introduced by Anzalone et. al by in situ melting of the silicon substrate once a certain 3C thickness was reached, eliminating the influence of TEC during the cool-down [124,125]. After the silicon removal, it is also possible to further increase the growth temperature, enabling growth of a few 100 μm-thick freestanding 3C-SiC wafers with reasonable growth rates and diameters up to 150 mm. While the 3C-SiC heteroepitaxy on Si is mostly performed above

1300 °C and near to the melting point of Si [126], low temperature CVD between 1000 °C and 1200 °C represents a different approach for the growth of 3C SiC on Si [127,128]. To compensate for the lower temperatures resulting in decreased growth rates, chlorine is often added either in the form of HCl or chlorine-containing precursors [129]. Although the quality of the achieved layers is not the same as for high temperature processes yet, the low temperatures open up the route for the integration into existing CMOS fabrication technology.

4.2.2. Heteroepitaxial Deposition on Hexagonal Substrates 4H SiC (CVD and PVT)

Additionally to CVD heteroepitaxy on Si, cubic SiC can be grown heteroepitaxially on hexagonal SiC substrates in either a CVD or physical vapour transport (PVT) setup. Using SiC as starting material enables higher growth temperature of up to 2000 °C and reduces the lattice mismatch as well as the difference in thermal expansion between substrate and epitaxial layer. For both cases (CVD and PVT), a (0001) orientated SiC substrate is usually used as a starting point, which will result in a (111) orientated 3C SiC top layer by "pseudomorph growth" [130]. During homoepitaxial growth of hexagonal SiC polytypes "step-controlled epitaxy" is preferred to ensure stable reproduction of the substrate polytype during growth [131–133]. Conditions favoring this growth regime are a low supersaturation at the growth surface as well as high off-axis angles (3–8°) resulting in a high number of atomic growth steps. The high step density and small terrace width, respectively, lead to a migration of adatoms towards step edges where they reproduce the substrate polytype during incorporation into the crystal. For the nucleation of 3C SiC on the hexagonal substrate, the growth conditions have to be tweaked from the nominal step flow towards the formation of 2D nucleation on the growth terraces. This growth regime can be promoted if the probability of adatoms reaching a growth step is reduced by either lowering of the growth temperature, increasing the terrace width by growth on "on-axis" substrates or increased supersaturation. A schematic scheme of the two growth modes are depicted in Figure 10.

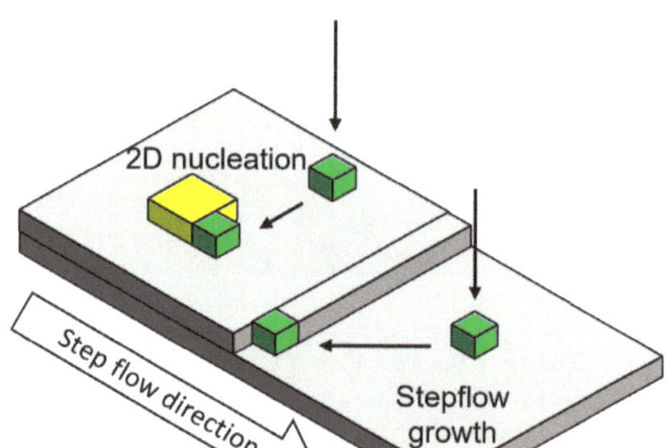

Figure 10. Scheme of two different growth schemes for the epitaxial growth. On the lower terrace, migration of the adatom at the edge of a growth step occurs, reproducing the substrate polytype. On the upper terrace, 2D nucleation takes place, which can lead to a switch in polytype.

Various studies can be found in the literature [134–138] working on the heteroepitaxial growth of 3C SiC on (0001) orientated hexagonal SiC up to a thickness of 2.5 mm using the sublimation epitaxy (SE) growth technique. They used a different off-axis substrate, on which an on axis facet will be formed at the initial growth stage, acting as a preferential nucleation side for the cubic polytype. Starting from this side, the 3C will laterally expand due to step flow growth covering the hexagonal substrate completely. The thickness necessary for a complete coverage is thereby predetermined by the use of the cut angle of the substrate and will increase for higher off-cuts. The SE setup creates a high supersaturation based on a small distance between source and seed and high temperature gradients, enhancing the polytype change towards 3C SiC. Additionally, the growth temperature is kept under 2000 °C in combination with an Si rich gas phase, which further stabilizes the cubic polytype. However, the nucleation of 3C SiC on hexagonal SiC polytypes is usually accompanied by the formation of double position boundaries (DPBs). During the nucleation of 3C SiC on hexagonal polytypes, the stacking sequence will be continued thermodynamically, dictated based on the underlying two bilayers [139]. Thus, if growth steps are present on the hexagonal substrate, 3C Seeds with varying stacking sequences will be formed on the different terraces, leading to defective merging during the lateral expansion of the 3C SiC, resulting in DPBs and SFs. Neudeck et al. [139] have presented a process to effectively eliminate the formation of DBPs during heteroepitaxial 3C SiC growth on 4H SiC substrates. They first produced a mesa-like structure by dry etching of (0001) orientated on-axis 4H SiC wafers. Afterwards, they used a homoepitaxial CVD growth process to completely remove atomistic growth steps from the mesas, followed by decreasing the temperature by 50–200 °C to initiate island growth of the cubic polytype. The lack of growth steps leads to the formation of 3C SiC seeds with the identical stacking sequence. Therefore, no DPBS will be formed during the merging of the 3C SiC seeds during lateral expansion. Note: The maximum size of the mesas was limited to 400 × 400 mm^2 and DPB-free mesas can only be obtained if threading screw dislocations (TSD) are absent in the 4H substrate underneath the mesas. TSDs will act as an infinite source of growth steps, inevitably leading to the formation of stacking defects.

4.2.3. Sublimation Growth

Different from the heteroepitaxial growth methods presented above, seeding material for the homoepitaxial growth of 3C SiC is not widely available yet. However, remarkable progress has been made on this topic in recent years. Schuh et al. [140] developed a transfer process to fabricate high-temperature stable 3C seeding stacks. Starting from 3C SiC grown heteroepitaxially on Si, the substrate was removed by wet chemical etching and the remaining free-standing 3C SiC was bonded to a polycrystalline SiC carrier using a carbon contain glue. By eliminating the low melting point of Si (1419 °C), the seeding stacks are suitable for the use in high-temperature sublimation growth processes, such as close space PVT (CS-PVT). Further information regarding the CS PVT setup can be found in [141] and chapter 5 of [142].

Using CS PVT and the mentioned seeding stacks, free-standing 3C SiC samples with diameters up to 100 mm and a thickness up to 3 mm could be realized [143,144]. The progress for the increase of sample diameter is depicted in Figure 11. Although the large area samples still suffer from cracking of the thin 3C SiC film during the removal of the CVD silicon substrate, the grown crystals show a high material quality, proved by the low full width half maximum (FWHM) of X-ray diffraction (XRD) measurements of the (002) reflex, which were as low as 140 arcsec. Note: Although a lot of research has been conducted to improve the material quality, the FWHM XRD rocking curve values for 3C SiC are still higher compared to 4H SiC where values in the range of 10 arcsec can be found in the literature [145,146]. Besides low FWHM values, it could be shown that the SF density as well as the overall stress in sublimation-grown 3C SiC crystal could be decreased compared to the heteroepitaxial-grown material on Si [143,147]. Based on the work of Anzalone et al. [124,125], which was already mentioned previously, free-standing, homoepitaxial-grown 3C SiC samples have become available. Although the material still has some setbacks, e.g., wafer bow, it is suitable as a starting point for the sublimation growth of 3C-SiC. Similar to the described seeding stacks, the homoepitaxial-grown seeds allow the growth of DBP-free material, as nucleating seeds will all have the same stacking sequence contrary to the heteroepitaxial growth on hexagonal SiC. However, compared to the seeding stacks, no transfer process is necessary for the homoepitaxial seeds, reducing the cracking probability and, consequently, increasing the up-scale potential. In fact, the first crack-free, free-standing large-scale 3C SiC crystals up to 650 μm thickness could be produced using this homoepitaxial seeding material, as can be seen in Figure 12 [148]. Further, Schöler et al. [149]) reported an overgrowth mechanism for protrusion defects using homoepitaxial seeds with an 4° off-cut towards the [100] direction. Protrusions represent a setback that hinders the growth of bulk material up to several millimeters in thickness. This three-dimensional defect forms close to the interface between silicon and 3C SiC during heteroepitaxial growth on silicon and can therefore be viewed as intrinsic for the used seeding material. During further growth, this defect will latterly increase with increasing layer thickness and distort the material quality. Inevitably, the surface of grown crystal will be covered by protrusions at some point during the growth, depending on the defect density. Although efforts have been made to reduce the protrusion density [111,150], this problem is far from being solved, as complete elimination of the defect is required for real bulk growth with boules thicknesses up to 10 mm or more.

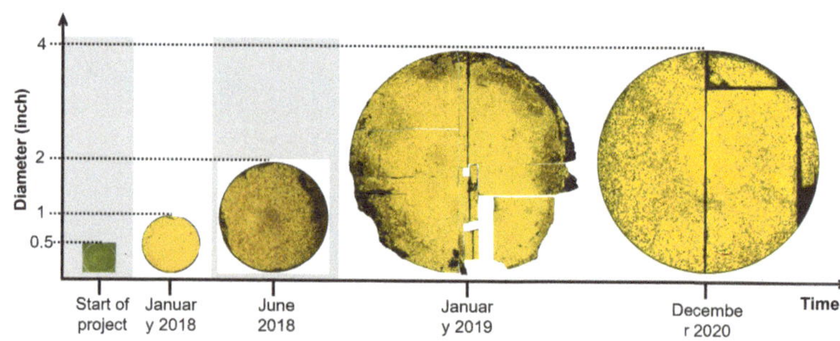

Figure 11. Evolution of diameters for bulk 3C-SiC crystals grown by sublimation growth during CHALLENGE project with indicated timeline. With permission from [145].

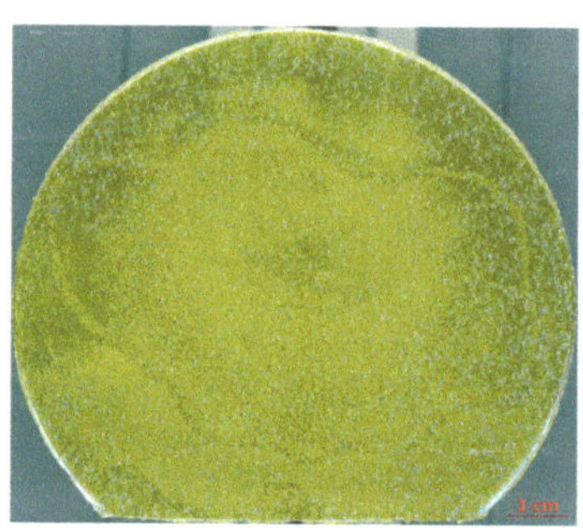

Figure 12. High resolution scan of free-standing 3C-SiC crystal grown by CS PVT on homoepitaxial seeding layer with a diameter of 92 mm and a thickness of 650 μm. With permission from [148].

4.3. *4H Silicon Carbide*

4.3.1. Bulk Growth of 4H SiC

The growth of bulk 4H SiC is mainly carried out by utilizing the physical vapour transport (PVT) method. Demonstrated in 1978 by Tairov and Tsvetkov [151], PVT enabled large-diameter and gas-phase single crystalline growth of hexagonal SiC, and is employed to grow crystals of up to 200 mm in diameter (see Figures 13 and 14). Other solution growth methods, such as top-seeded solution growth (TSSG) of SiC, are also conceivable. However, this method is held back by the low solubility of carbon in a silicon melt. According to the phase diagram, SiC exhibits a peritectic decomposition at approximately 2830 °C into carbon and an Si-rich Si-C solution [152]. It has to be mentioned that crystal growth from the melt is thermodynamically advantageous compared to vapour phase growth and very high-quality SiC crystals have been demonstrated with TSSG [153]. However, the obstacles of the non-stoichiometric melt composition, combined with the early availability and comparative ease of the implementation of PVT growth, led to an almost exclusive use of PVT for the production of commercial SiC substrates.

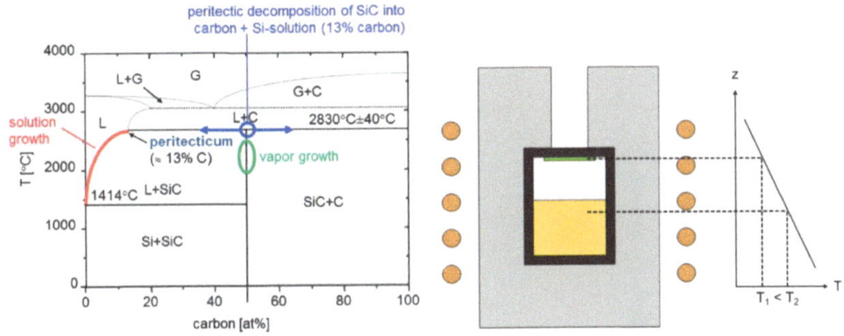

(a) **(b)**

Figure 13. (**a**) Phase diagram of Si and C. (**b**) Schematic setup of a typical PVT growth cell and isolation. With permission from [154].

Figure 14. 100 mm (**left**) and 150 mm (**right**) 4H-SiC crystal.

Crystal growth by PVT is taking place by inductively heating a graphite crucible to temperatures above 2000 °C. While the high temperatures of the PVT process allow good-quality crystals to grow, it also prevents any kind of CMOS compatibility. Inside the crucible, a SiC source, predominantly a powder, is placed below a single crystalline seed. A temperature gradient between the source powder and the seed ensures that mass transport will take place once sufficient growth temperatures are reached. The SiC powder source will sublimate and the resulting SiC gas species will recrystallize at the slightly colder seed. The mass transport, and therefore the growth rate, can be adjusted by varying the growth temperature in general, the axial temperature gradient inside the crucible or the growth pressure. All materials inside the growth chamber of the PVT setup need to be stable at high growth temperatures without breaking down in order to prevent changing growth conditions during the growth run or unintentional doping of the growing crystal. For this reason, graphite parts and isolations are utilized combined with a gas composition of argon and, if necessary, nitrogen for n-type doping. P-type doping with aluminium is possible as well, as demonstrated by Wellmann et al. [155]. The crystal growth is started by heating up to growth temperatures and the subsequent lowering of the pressure. The growth rate lies in the range of 50 to 500 µm/h but, depending on the growth cell design, can be set to rates up to 1 mm/h. Since hexagonal polytypes of SiC are grown at much higher temperatures than the cubic polytype, the crystal quality also tends to be of a higher quality with less structural defects. A more extensive description of the mechanics of PVT growth can be read elsewhere [156,157].

4.3.2. Defects in Hexagonal SiC

The main challenge in PVT growth lies in the minimization of defects present in the crystal lattice. The diminishing effect of defects on the performance of SiC-based devices has been reported extensively. Micropipes (MP) can lead to breakdowns of p-n junctions [158,159], high densities of threading screw dislocations (TSD) have been correlated to lower breakdown voltages of 4H-SiC rectifiers and basal plane dislocations (BPD) are reported to increase the leakage current in MOSFETs [160] or JFETs [161]. While threading edge dislocations (TED) are also reported to reduce the breakdown voltage of Schottky devices [162], they have the least impact of the aforementioned defect types. Defects can either be generated due to process conditions or are inherited from the seed. MPs, TSDs and TEDs lie parallel to the growth direction while BPDs lie within the basal plane perpendicular to the growth direction. Dislocations can be revealed by selective etching with molten KOH at 500–520 °C for 5–10 min [163]. Only the Si-terminated face (so the (0001)-plane) is etched in an anisotropic fashion. The (0001)-plane or C-face is etched with an etching rate four times as fast and does not reveal any etch pits [164].

Figure 15 depicts a typical SiC wafer surface etched with molten KOH. All main defect types are shown. The BPD exhibits a characteristic elongated etch pit since the dislocation line lies perpendicular to the growth direction. BPDs are only visible in KOH-etched samples if there is a slight off-axis angle in respect to the (0001)-plane. TEDs are expressed by a small circular etch pit while TSD etch pits have a more pronounced hexagonal shape. In Figure 15, you can see several sizes of TSD etch pits. The reason for that lies in the different values the burgers vectors of TSDs can assume. It was demonstrated that TSDs exist with burgers vectors $b \leq 3c$ [165]. The different sizes of the TSDs in the depicted KOH-image most likely correlate to different amounts of b. The largest etch pits belong to MPs. Since MPs are essentially TSDs with an open core and burgers vectors $b \geq 3c$ [166], the etch pit looks up-scaled but similar. The overlap between MPs and TSDs at 3c means that, without directly observing the open core of MPs, it is very difficult to differentiate between a 3c MP and a 3c TSD in a KOH etching image. Other defect characterization methods worth mentioning are photoluminescent mapping, X-ray topography or Raman spectroscopy. A more extensive overview can be found in [167].

MPs are formed by carbon inclusions [168], adsorbing onto the growth interface during the process and inhibiting the step-flow, polytype switches during the growth [169] and insufficient backside protection against sublimation of the crystal [170,171]. These issues are largely resolved by now, demonstrated by growing MP-free 4H-SiC single crystals of up to 200 mm diameter [172]. The main focus currently lies in the reduction of the TSD and BPD density. TSDs are known to be formed due to the formation of independent growth islands during the initial seeding phase [173]. This can be remedied by utilizing off-axis seeds. BPDs, on the other hand, are induced by stress in the crystal, either during or after the growth is concluded, since deformation in hexagonal SiC polytypes takes place in the $<11\bar{2}0>\{0001\}$ slip system. As a conclusion, the stress present in the crystal and the BPD density is closely connected. To reduce the BPD density, the different sources of stress have to be considered. One is the difference in the coefficient of thermal expansion (CTE) of the different materials utilized in PVT growth. A crystal can be considered to be growing in a stress-free manner at growth temperatures since the thermal energy will enable it to include dislocations to adapt to present strain immediately. If the crystal is firmly connected to graphite parts, such as the seed holder or the crucible wall, during cool-down, the different CTEs of SiC and graphite will lead to stress and the generation of BPDs before reaching the ductile-brittle transition temperature of 1050 °C [174]. However, even if all contact to graphite is prevented, the temperature gradients inside the crystal before cooling down will induce stress once room temperature is reached [175,176]. This is also true for the radial temperature gradient. It was demonstrated numerically and experimentally that, from the shoulder region of a crystal, BPD arrays form and propagate into the crystal [177,178]. Therefore, low radial and axial gradients are preferred, without either inducing a concave growth interface or reducing the mass transport between source and seed too much. The control of the temperature gradients becomes even more important if bigger diameters of crystals, such as 200 mm, are to be grown since the total difference in temperature between the center of the crystal and its edge increases. Additionally, nitrogen doping in SiC modifies the CTE of hexagonal SiC [179]. Due to the nature of PVT growth, during the seeding phase, adsorbed nitrogen gas species can release from the graphite isolations, which subsequently leads to a sharp increase of doping for the first few µm of crystal growth. This, in turn, leads to a high amount of stress during the cool-down phase [180]. A controlled gradual increase of the nitrogen gas flux during the start of growth will prevent the inhomogeneous doping and therefore avoid excessive BPD formation. In addition, the utilized graphite parts should be of high purity, in the range of 6N, optimally combined with a purging step prior to growing. This is especially important if semi-insulating SiC for the novel photonic application in the topic of quantum information is desired, elevating the need for purity into magnitudes of 7N to 8N.

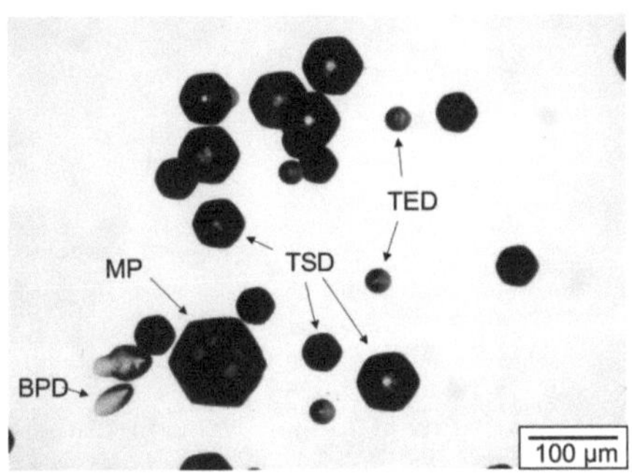

Figure 15. Microscopic image of typical etch pits in a 4H/6H-SiC wafer, etched with KOH.

4.3.3. X-ray Imaging

The high temperatures during the PVT process prevent a direct observation of the growing crystal. PVT setups are mostly limited to measure the temperatures above and below the growth cell with optical pyrometers through thin channels in the graphite isolation. However, it is possible to utilize 2D X-ray imaging to approximate the in situ growth rate and powder consumption [181–183], enabling the fine-tuning of parameters, such as the thermal field and consumption rate of the SiC source. For more specific inquiries of the growth process, advanced 3D imaging can be employed. Such sophisticated systems are not commercially viable due to the technical complexities; however, in an R&D environment, they have been proven to be incredibly valuable in considering problems such as the evolution of the crystal's facet during growth [184], the specific morphology of the powder source for the determination of dynamic source material properties [185], as well as the impact of the curvature of the growth interface on the defect distribution in the resulting crystal [186].

4.3.4. Numerical Modeling

Besides X-ray imaging, numerical modeling is utilized to further characterize the growth cell. This tool can be used to adjust a wide range of parameters, such as the shape of the temperature field inside the crucible, without having to perform time- and material-consuming test runs [185,187,188]. Figure 16 depicts a typical modeling result. To obtain a high accuracy of the modeling results, the main task is a precise knowledge of the material's properties over a wide range of temperature. Laser flash analysis can be employed to obtain the temperature-dependent thermal conductivity while dilatometry is useful to characterize the thermal expansion. The temperature measurements provided by pyrometers can be used to validate the model. Further research of the SiC material system allowed to model the mass flow inside the crucible during a process [187,189,190]. In [190], ^{13}C was utilized in the powder source, combined with X-ray imaging to track the mass flow during a PVT growth run and the results were subsequently compared with the numerical calculations. The growth kinetics within the growth cell and the stress acting on the growing crystal were investigated by several workgroups [180,191,192], followed by the assessment of the formation and movement of dislocations [136,177,193,194]. State-of-the-art modeling of the PVT growth cell allows the accurate depiction of the conditions present during the growth process, which would otherwise not be obtainable.

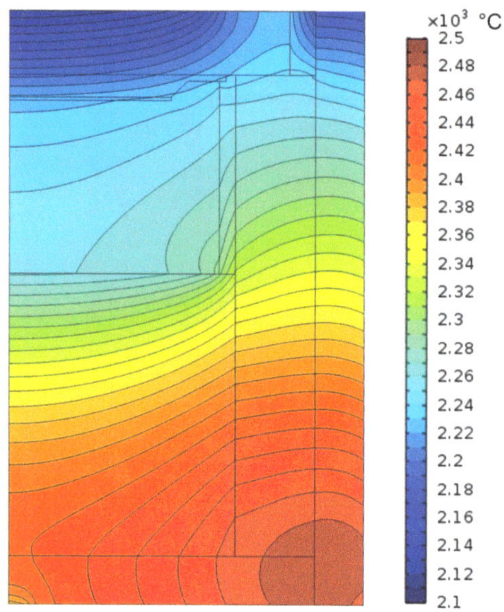

Figure 16. Calculated temperature field of a typical growth cell in a PVT reactor.

4.3.5. CVD Growth of 4H-SiC

Like the cubic polytype, 4H-SiC can also be epitaxially grown employing the CVD method, as was first reported by Matsunami and Kimoto [195]. Since the 3C polytype is thermodynamically more stable than the hexagonal polytypes at the typical growth temperatures of the CVD process, homoepitaxial growth of 4H-SiC needs off-cut substrates to prevent polytype switches. As long as the step-flow growth mode, as described by the BCF model, is retained [196], the polytype of the substrate can be reproduced even though the thermodynamic conditions for hexagonal SiC are unfavorable at lower temperatures [197]. This holds true as long as there are no growth islands on large-area terraces. One important advantage of the CVD-growth of 4H-SiC layers is the conversion of BPDs to TEDs through image-force effects present at lower off-cut angles, such as 2 to 4° [198,199]. To some degree, this conversion is also happening in PVT-growth, although there, it is much less controllable. The general mechanics of homoepitaxial CVD growth are the same as for the heteroepitaxial growth of 3C on Si, described in the previous section. Special care has to be taken to prevent disruptions of the step-flow caused by side reactions during the growth process, such as condensation of Si atoms and resulting Si droplets. While a lower off-cut angle of the substrate increases the image-force and, therefore, the conversion of BPDs into TEDs, it also increases the size of terraces and the probability of 3C-SiC islands. To prevent this, the diffusion length of an adsorbed Si or C atom on the substrate should at least reach half of the terrace width. The main parameters to vary are the growth temperature and the gas fluxes of the precursor gases silane and propane. Compared to PVT-growth, the growth rate seldom reaches values above 50 μm/h, making CVD-growth unsuitable for bulk growth of SiC. However, the process temperature can be set to values below 1200 °C, making a CMOS compatibility possible.

5. Conclusion and Perspectives

The exceptional optical properties of bulk SiC will be enhanced when light is confined in nanoscale waveguides, so that the devices become more compact and energy efficient. Broadly speaking, SiC has been used in the photonic field in such cases as being a substrate for high-efficiency GaN-based LED because of its small lattice mismatch to GaN, emitting light by doping donar-acceptor pairs and forming porous structures [200,201]. In this review paper, a novel photonic application of SiC is presented. Although this field is very young, about 10 years old, many breakthroughs have been achieved, thanks to the already mature SiC growth and fabrication technologies. Compared to the longer-existing material platforms for photonic integrated circuits (Si, SiO_2, SiN, $LiNbO_3$, AlGaAs, etc.), SiC is outperforming on both its nonlinear optical properties and its potential as a perfect candidate for a single photon source for future quantum photonic integrated circuits. SiCOI stacks with a polytype of 4H, 3C and amorphous have been demonstrated by different formation methods, and high-quality-factor (>1 million) microring resonators have been reported. A lot of passive and active devices are demonstrated, such as beam splitter, polarization beam splitter, modulator, frequency comb, etc [23,38,63,74].

From a loss reduction point of view, any defects causing absorption and scattering of light at the device working wavelength should be eliminated. However, there are a bunch of point defects in SiC which are optically addressable and could be potentially used for a single photon source, a key device for quantum technology. Integrating SiC single photon source, monolithically, with other building blocks enables SiCOI a game-changer for the future of quantum PIC [202].

But, before SiC really plays a crucial role in QPIC, further research and development are needed:

(1) Low-loss SiCOI stacks are widely available.
(2) The insertion loss of the chip, including propagation loss and coupling should be less than 3 dB.
(3) More SiC-based quantum devices should be demonstrated, such as quantum memory, detectors, etc.
(4) SiC color centers should be further explored and prepared as a competitive candidate for a single photon source for real applications in communication, computing, etc.

6. Impacts of Novel Photonic Applications of SiC

The research in silicon carbide started to have strong momentum in the 1990s. That decade had a heavy focus in research and in Europe several EU projects were initiated. The first ones were mainly in bulk growth and included some epitaxy and device development. Epitaxy and device challenges were heavily related to defects in research-grade wafers, and epitaxy recipes and issues such as stable electrical contacts were not in place. Test devices suffered from structural variations within wafer areas, as well as between wafers. The first European wafer suppliers emerged 20–25 years ago and slowly substrates became more available for device development. The market started to grow slowly, firstly by the sale of materials. It was only when devices became available that the market started to grow faster. The SiC diodes and power electronics pushed the market, and also started to require more wafers. As a consequence, the wafer prices have steadily decreased and volume manufacturing is now established. Today, many market players are entering and the SiC device market will grow to billions of US dollars annually. The journey from research to a strong, growing SiC market has taken about 30 years.

The efforts in the research decade created a strong momentum in Europe. Researchers built their competence and gained network and personal relations. While conducting present projects, new ideas emerged through the exchange between researchers. The fluorescent silicon carbide for a new white LED was one of the European avenues which gathered a cluster of research groups in the year 2010, with teams from Denmark, Germany and Sweden, which now present this review article. They have continued to collaborate and propose new projects with other partners having complementary expertise. In such a way, silicon carbide research expanded to other avenues and started to grow to applications other than the original ones that were related to power devices. The European efforts in SiC from the 1990s have grown momentum.

The next phase of momentum is given by impact creation. Projects are typically described by complementary partners and value chains. The motivations are given by technical advantages and better performances. Increasingly, there are aims like the Green Deal, Sustainable Development Goals (SDGs), and others which have a long-term aspect. Even though those are reasonable motivations, there should be sooner goals which show synergies. This is where regional development of Smart Specialisation Strategies (S3) comes in. It is a place-based innovation policy concept in which regional priorities are supported. This is achieved through the entrepreneurial discovery process in innovative sectors, fields or technologies. It applies a bottom-up approach to find and support regional scientific and technological development. The S3 is, in simplified terms, described as ways to create more economical growth in European regions. By finding areas of strength and specializing in more efforts such as this one, there will be a sustainable way to increase economic growth. This is basically achieved by finding partners in regions and collaboration in regions which are complementary. It, in fact, describes various ways to have value chains. Thus, value chains are common in both technology and smart specialization motivations and implementation. The S3 was introduced about ten years ago. To date, more than 120 Smart Specialization Strategies are developed in EU regions and Member States. Now comes Sustainable Smart Specialization Strategies (S4). The S4 has an emphasis on the importance of taking a long-term view of the development of a region. It should not just be as an adaptive path, rather, it should be a proactive path.

The SiComb project is an H2020 EU-FET Open project that explores the photonic application using SiC, as described in the technical part of this paper. The EU FET Open, and EIC Pathfinder, as denoted in the Horizon Europe Programme, are projects that present ideas with potential radical technology leaps, looking forward 10–20 years. These are early stages of research with potential for high gain, if the research and technology challenges can be managed.

Clearly, there can be synergies and proactive development by aligning technology and regional development and their motivations to gather stakeholders. Creating exchange between different stakeholders, such as researchers, regional development actors, policy makers, citizens, and other relevant players will produce the momentum which aligns the development of smart specialization and technology for society. The common overall aim of sustainable approaches for SDGs and the Green Deal will be more efficient by considering such joint impact creation.

Author Contributions: Conceptualization, H.O. and P.W.; validation, P.W. and D.C.; investigation, H.O., X.S., Y.L., M.K., J.S. and V.T.; writing—original draft preparation, H.O., X.S., Y.L., M.K., J.S., V.T., M.S., P.W. and D.C.; writing—review and editing, H.O., X.S., Y.L. and D.C.; visualization, H.O., X.S., Y.L., M.K. and J.S.; supervision, H.O., P.W. and D.C.; funding acquisition, H.O. and P.W; utilisation and innovation, M.S. All authors have read and agreed to the published version of the manuscript.

Funding: This research was funded by European Union's Horizon 2020 FET Open (SiComb, No. 899679) and VILLUM FONDEN (VIL50293).

Institutional Review Board Statement: Not applicable.

Informed Consent Statement: Not applicable.

Data Availability Statement: The data presented in this study are available on request from the corresponding author.

Conflicts of Interest: The authors declare no conflict of interest.

References

1. Dinu, M.; Quochi, F.; Garcia, H. Third-order nonlinearities in silicon at telecom wavelengths. *Appl. Phys. Lett.* **2003**, *82*, 2954–2956. [CrossRef]
2. Griffith, A.G.; Lau, R.K.; Cardenas, J.; Okawachi, Y.; Mohanty, A.; Fain, R.; Lee, Y.H.D.; Yu, M.; Phare, C.T.; Poitras, C.B.; et al. Silicon-chip mid-infrared frequency comb generation. *Nat. Commun.* **2015**, *6*, 6299. [CrossRef] [PubMed]
3. Guo, K.; Shi, X.; Wang, X.; Yang, J.; Ding, Y.; Ou, H.; Zhao, Y. Generation rate scaling: The quality factor optimization of microring resonators for photon-pair sources. *Photonics Res.* **2018**, *6*, 587–596. [CrossRef]
4. Guo, K.; Yang, L.; Shi, X.; Liu, X.; Cao, Y.; Zhang, J.; Wang, X.; Yang, J.; Ou, H.; Zhao, Y. Nonclassical optical bistability and resonance-locked regime of photon-pair sources using silicon microring resonator. *Phys. Rev. Appl.* **2019**, *11*, 034007. [CrossRef]
5. Shi, X.; Guo, K.; Christensen, J.B.; Castaneda, M.A.U.; Liu, X.; Ou, H.; Rottwitt, K. Multichannel photon-pair generation with strong and uniform spectral correlation in a silicon microring resonator. *Phys. Rev. Appl.* **2019**, *12*, 034053. [CrossRef]
6. Kippenberg, T.J.A. *Nonlinear Optics in Ultra-High-Q Whispering-Gallery Optical Microcavities*; California Institute of Technology: Pasadena, CA, USA, 2004.
7. Ou, H. Different index contrast silica-on-silicon waveguides by PECVD. *Electron. Lett.* **2003**, *39*, 1. [CrossRef]
8. Xuan, Y.; Liu, Y.; Xue, X.; Wang, P.H.; Wang, J.; Niu, B.; Han, K.; Teng, M.; Leaird, D.E.; Weiner, A.M.; et al. High-Q silicon nitride microresonator for low power frequency comb initiation at normal dispersion regime. In Proceedings of the CLEO: Science and Innovations, San Jose, CA, USA, 10–15 May 2015; Optica Publishing Group: Washington, DC, USA, 2015.
9. Wang, L.; Chang, L.; Volet, N.; Pfeiffer, M.H.; Zervas, M.; Guo, H.; Kippenberg, T.J.; Bowers, J.E. Frequency comb generation in the green using silicon nitride microresonators. *Laser Photonics Rev.* **2016**, *10*, 631–638. [CrossRef]
10. Philipp, H.T.; Andersen, K.N.; Svendsen, W.; Ou, H. Amorphous silicon rich silicon nitride optical waveguides for high density integrated optics. *Electron. Lett.* **2004**, *40*, 1. [CrossRef]
11. Razzari, L.; Duchesne, D.; Ferrera, M.; Morandotti, R.; Chu, S.; Little, B.E.; Moss, D.J. CMOS-compatible integrated optical hyper-parametric oscillator. *Nat. Photonics* **2010**, *4*, 41–45. [CrossRef]
12. Martin, A.; Combrié, S.; de Rossi, A.; Beaudoin, G.; Sagnes, I.; Raineri, F. Nonlinear gallium phosphide nanoscale photonics. *Photonics Res.* **2018**, *6*, B43–B49. [CrossRef]
13. Wilson, D.J.; Schneider, K.; Hönl, S.; Anderson, M.; Baumgartner, Y.; Czornomaz, L.; Kippenberg, T.J.; Seidler, P. Integrated gallium phosphide nonlinear photonics. *Nat. Photonics* **2020**, *14*, 57–62. [CrossRef]
14. Golovan, L.A.; Mel'nikov, V.A.; Konorov, S.O.; Fedotov, A.B.; Gavrilov, S.A.; Zheltikov, A.M.; Kashkarov, P.K.; Timoshenko, V.Y.; Petrov, G.I.; Li, L.; et al. Efficient second-harmonic generation by scattering from porous gallium phosphide. *J. Exp. Theor. Phys. Lett.* **2003**, *78*, 193–197. [CrossRef]
15. Jung, H.; Xiong, C.; Fong, K.Y.; Zhang, X.; Tang, H.X. Optical frequency comb generation from aluminum nitride microring resonator. *Opt. Lett.* **2013**, *38*, 2810–2813. [CrossRef] [PubMed]
16. Jung, H.; Stoll, R.; Guo, X.; Fischer, D.; Tang, H.X. Green, red, and IR frequency comb line generation from single IR pump in AlN microring resonator. *Optica* **2014**, *1*, 396–399. [CrossRef]
17. Hausmann, B.; Bulu, I.; Venkataraman, V.; Deotare, P.; Lončar, M. Diamond nonlinear photonics. *Nat. Photonics* **2014**, *8*, 369–374. [CrossRef]
18. Pu, M.; Ottaviano, L.; Semenova, E.; Yvind, K. Efficient frequency comb generation in AlGaAs-on-insulator. *Optica* **2016**, *3*, 823–826. [CrossRef]
19. Marino, G.; Gigli, C.; Rocco, D.; Lemaitre, A.; Favero, I.; De Angelis, C.; Leo, G. Zero-order second harmonic generation from AlGaAs-on-insulator metasurfaces. *ACS Photonics* **2019**, *6*, 1226–1231. [CrossRef]
20. Zhu, D.; Shao, L.; Yu, M.; Cheng, R.; Desiatov, B.; Xin, C.; Hu, Y.; Holzgrafe, J.; Ghosh, S.; Shams-Ansari, A.; et al. Integrated photonics on thin-film lithium niobate. *Adv. Opt. Photonics* **2021**, *13*, 242–352.
21. Qin, H.; Shi, X.; Ou, H. Exceptional points at bound states in the continuum in photonic integrated circuits. *Nanophotonics* **2022**, *11*, 4909–4917. [CrossRef]
22. Liang, X.; Fu, L.; Yu, Q.; Xue, Z.; Shi, X.; Lu, Y.; Chen, H.; Zhang, B.; Luo, Y.; Hu, Q.; et al. Efficient and Broadband Trident Spot-Size Convertor for Thin-Film Lithium Niobate Integrated Device. *IEEE Photonics Technol. Lett.* **2023**, *35*, 35–38. [CrossRef]
23. Shi, X.; Fan, W.; Lu, Y.; Hansen, A.K.; Chi, M.; Yi, A.; Ou, X.; Rottwitt, K.; Ou, H. Polarization and spatial mode dependent four-wave mixing in a 4H-silicon carbide microring resonator. *APL Photonics* **2021**, *6*, 076106. [CrossRef]
24. Shi, X.; Zhang, J.; Fan, W.; Lu, Y.; Peng, N.; Rottwitt, K.; Ou, H. Compact low-birefringence polarization beam splitter using vertical-dual-slot waveguides in silicon carbide integrated platforms. *Photonics Res.* **2022**, *10*, A8–A13.
25. Yi, A.; Zheng, Y.; Huang, H.; Lin, J.; Yan, Y.; You, T.; Huang, K.; Zhang, S.; Shen, C.; Zhou, M.; et al. Wafer-scale 4H-silicon carbide-on-insulator (4H–SiCOI) platform for nonlinear integrated optical devices. *Opt. Mater.* **2020**, *107*, 109990.
26. Castelletto, S.; Peruzzo, A.; Bonato, C.; Johnson, B.C.; Radulaski, M.; Ou, H.; Kaiser, F.; Wrachtrup, J. Silicon Carbide Photonics Bridging Quantum Technology. *ACS Photonics* **2022**, *9*, 1434–1457.
27. Lukin, D.M.; Guidry, M.A.; Vučković, J. Integrated quantum photonics with silicon carbide: Challenges and prospects. *PRX Quantum* **2020**, *1*, 020102.

28. Song, B.S.; Yamada, S.; Asano, T.; Noda, S. Demonstration of two-dimensional photonic crystals based on silicon carbide. *Opt. Express* **2011**, *19*, 11084–11089. [CrossRef]
29. Yamada, S.; Song, B.S.; Asano, T.; Noda, S. Experimental investigation of thermo-optic effects in SiC and Si photonic crystal nanocavities. *Opt. Lett.* **2011**, *36*, 3981–3983. [CrossRef]
30. Yamada, S.; Song, B.S.; Asano, T.; Noda, S. Silicon carbide-based photonic crystal nanocavities for ultra-broadband operation from infrared to visible wavelengths. *Appl. Phys. Lett.* **2011**, *99*, 201102. [CrossRef]
31. Yamada, S.; Song, B.S.; Upham, J.; Asano, T.; Tanaka, Y.; Noda, S. Suppression of multiple photon absorption in a SiC photonic crystal nanocavity operating at 1.55 μm. *Opt. Express* **2012**, *20*, 14789–14796. [CrossRef]
32. Yamada, S.; Song, B.S.; Jeon, S.; Upham, J.; Tanaka, Y.; Asano, T.; Noda, S. Second-harmonic generation in a silicon-carbide-based photonic crystal nanocavity. *Opt. Lett.* **2014**, *39*, 1768–1771. [CrossRef]
33. Martini, F.; Politi, A. Four wave mixing in 3C SiC ring resonators. *Appl. Phys. Lett.* **2018**, *112*, 251110. [CrossRef]
34. Fan, T.; Moradinejad, H.; Wu, X.; Eftekhar, A.A.; Adibi, A. High-Q integrated photonic microresonators on 3C-SiC-on-insulator (SiCOI) platform. *Opt. Express* **2018**, *26*, 25814–25826. [CrossRef]
35. Xing, P.; Ma, D.; Ooi, K.J.; Choi, J.W.; Agarwal, A.M.; Tan, D. CMOS-compatible PECVD silicon carbide platform for linear and nonlinear optics. *ACS Photonics* **2019**, *6*, 1162–1167.
36. Zheng, Y.; Pu, M.; Yi, A.; Chang, B.; You, T.; Huang, K.; Kamel, A.N.; Henriksen, M.R.; Jørgensen, A.A.; Ou, X.; et al. High-quality factor, high-confinement microring resonators in 4H-silicon carbide-on-insulator. *Opt. Express* **2019**, *27*, 13053–13060. [PubMed]
37. Cardenas, J.; Yu, M.; Okawachi, Y.; Poitras, C.B.; Lau, R.K.; Dutt, A.; Gaeta, A.L.; Lipson, M. Optical nonlinearities in high-confinement silicon carbide waveguides. *Opt. Lett.* **2015**, *40*, 4138–4141. [PubMed]
38. Guidry, M.A.; Yang, K.Y.; Lukin, D.M.; Markosyan, A.; Yang, J.; Fejer, M.M.; Vučković, J. Optical parametric oscillation in silicon carbide nanophotonics. *Optica* **2020**, *7*, 1139–1142.
39. Lu, X.; Lee, J.Y.; Feng, P.X.L.; Lin, Q. Silicon carbide microdisk resonator. *Opt. Lett.* **2013**, *38*, 1304–1306.
40. Cardenas, J.; Zhang, M.; Phare, C.T.; Shah, S.Y.; Poitras, C.B.; Guha, B.; Lipson, M. High q sic microresonators. *Opt. Express* **2013**, *21*, 16882–16887.
41. Radulaski, M.; Babinec, T.M.; Buckley, S.; Rundquist, A.; Provine, J.; Alassaad, K.; Ferro, G.; Vučković, J. Photonic crystal cavities in cubic (3C) polytype silicon carbide films. *Opt. Express* **2013**, *21*, 32623–32629.
42. Lu, X.; Lee, J.Y.; Feng, P.X.L.; Lin, Q. High Q silicon carbide microdisk resonator. *Appl. Phys. Lett.* **2014**, *104*, 181103.
43. Calusine, G.; Politi, A.; Awschalom, D.D. Silicon carbide photonic crystal cavities with integrated color centers. *Appl. Phys. Lett.* **2014**, *105*, 011123.
44. Lu, X.; Lee, J.Y.; Rogers, S.; Lin, Q. Optical Kerr nonlinearity in a high-Q silicon carbide microresonator. *Opt. Express* **2014**, *22*, 30826–30832. [PubMed]
45. Lee, J.Y.; Lu, X.; Lin, Q. High-Q silicon carbide photonic-crystal cavities. *Appl. Phys. Lett.* **2015**, *106*, 041106.
46. Bracher, D.O.; Hu, E.L. Fabrication of high-Q nanobeam photonic crystals in epitaxially grown 4H-SiC. *Nano Lett.* **2015**, *15*, 6202–6207.
47. Lohrmann, A.; Karle, T.J.; Sewani, V.K.; Laucht, A.; Bosi, M.; Negri, M.; Castelletto, S.; Prawer, S.; McCallum, J.C.; Johnson, B.C. Integration of single-photon emitters into 3C-SiC microdisk resonators. *ACS Photonics* **2017**, *4*, 462–468.
48. Bracher, D.O.; Zhang, X.; Hu, E.L. Selective Purcell enhancement of two closely linked zero-phonon transitions of a silicon carbide color center. *Proc. Natl. Acad. Sci. USA* **2017**, *114*, 4060–4065.
49. Martini, F.; Politi, A. Linear integrated optics in 3C silicon carbide. *Opt. Express* **2017**, *25*, 10735–10742. [CrossRef]
50. Song, B.S.; Asano, T.; Jeon, S.; Kim, H.; Chen, C.; Kang, D.D.; Noda, S. Ultrahigh-Q photonic crystal nanocavities based on 4H silicon carbide. *Optica* **2019**, *6*, 991–995.
51. Wu, X.; Fan, T.; Eftekhar, A.A.; Adibi, A. High-Q microresonators integrated with microheaters on a 3C-SiC-on-insulator platform. *Opt. Lett.* **2019**, *44*, 4941–4944.
52. Lukin, D.M.; Dory, C.; Guidry, M.A.; Yang, K.Y.; Mishra, S.D.; Trivedi, R.; Radulaski, M.; Sun, S.; Vercruysse, D.; Ahn, G.H.; et al. 4H-silicon-carbide-on-insulator for integrated quantum and nonlinear photonics. *Nat. Photonics* **2020**, *14*, 330–334.
53. Fan, T.; Wu, X.; Eftekhar, A.A.; Bosi, M.; Moradinejad, H.; Woods, E.V.; Adibi, A. High-quality integrated microdisk resonators in the visible-to-near-infrared wavelength range on a 3C-silicon carbide-on-insulator platform. *Opt. Lett.* **2020**, *45*, 153–156.
54. Crook, A.L.; Anderson, C.P.; Miao, K.C.; Bourassa, A.; Lee, H.; Bayliss, S.L.; Bracher, D.O.; Zhang, X.; Abe, H.; Ohshima, T.; et al. Purcell enhancement of a single silicon carbide color center with coherent spin control. *Nano Lett.* **2020**, *20*, 3427–3434. [PubMed]
55. Xing, P.; Ma, D.; Kimerling, L.C.; Agarwal, A.M.; Tan, D.T. High efficiency four wave mixing and optical bistability in amorphous silicon carbide ring resonators. *APL Photonics* **2020**, *5*, 076110. [CrossRef]
56. Wang, C.; Fang, Z.; Yi, A.; Yang, B.; Wang, Z.; Zhou, L.; Shen, C.; Zhu, Y.; Zhou, Y.; Bao, R.; et al. High-Q microresonators on 4H-silicon-carbide-on-insulator platform for nonlinear photonics. *Light. Sci. Appl.* **2021**, *10*, 139. [PubMed]
57. Cai, L.; Li, J.; Wang, R.; Li, Q. Octave-spanning microcomb generation in 4H-silicon-carbide-on-insulator photonics platform. *Photonics Res.* **2022**, *10*, 870–876.
58. Lu, Y.; Shi, X.; Chaussende, D.; Rottwitt, K.; Ou, H. Thermal-tuning of Amorphous Silicon Carbide Nanophotonic Devices. In Proceedings of the CLEO: Science and Innovations, San Jose, CA, USA, 15–20 May 2022; Optica Publishing Group: Washington, DC, USA, 2022.

59. Allioux, D.; Belarouci, A.; Hudson, D.; Magi, E.; Sinobad, M.; Beaudin, G.; Michon, A.; Singh, N.; Orobtchouk, R.; Grillet, C. Toward mid-infrared nonlinear optics applications of silicon carbide microdisks engineered by lateral under-etching. *Photonics Res.* **2018**, *6*, B74–B81.
60. Wang, C.; Shen, C.; Yi, A.; Yang, S.; Zhou, L.; Zhu, Y.; Huang, K.; Song, S.; Zhou, M.; Zhang, J.; et al. Visible and near-infrared microdisk resonators on a 4H-silicon-carbide-on-insulator platform. *Opt. Lett.* **2021**, *46*, 2952–2955.
61. Powell, K.; Shams-Ansari, A.; Desai, S.; Austin, M.; Deng, J.; Sinclair, N.; Lončar, M.; Yi, X. High-Q suspended optical resonators in 3C silicon carbide obtained by thermal annealing. *Opt. Express* **2020**, *28*, 4938–4949.
62. Shi, X.; Fan, W.; Lu, Y.; Yi, A.; Ou, X.; Rottwitt, K.; Ou, H. Thermal oxidation assisted chemical mechanical polishing for low-loss 4H-SiC integrated photonic devices. In Proceedings of the 47th Micro and Nano Engineering Conference, Turin, Italy, 20–23 September 2021.
63. Shi, X.; Lu, Y.; Peng, N.; Rottwitt, K.; Ou, H. High-performance polarization-independent beam splitters and MZI in silicon carbide integrated platforms for single-photon manipulation. *J. Light. Technol.* **2022**, *40*, 7626–7633.
64. Shi, X.; Lu, Y.; Ou, H. High-performance silicon carbide polarization beam splitting based on the asymmetric directional coupler for mode conversion. *Opt. Lett.* **2022**, *48*, 1–5.
65. Bogaerts, W.; De Heyn, P.; Van Vaerenbergh, T.; De Vos, K.; Kumar Selvaraja, S.; Claes, T.; Dumon, P.; Bienstman, P.; Van Thourhout, D.; Baets, R. Silicon microring resonators. *Laser Photonics Rev.* **2012**, *6*, 47–73.
66. Xiong, M.; Lei, L.; Ding, Y.; Huang, B.; Ou, H.; Peucheret, C.; Zhang, X. All-optical 10 Gb/s AND logic gate in a silicon microring resonator. *Opt. Express* **2013**, *21*, 25772–25779. [PubMed]
67. Xiong, M.; Ozolins, O.; Ding, Y.; Huang, B.; An, Y.; Ou, H.; Peucheret, C.; Zhang, X. Simultaneous RZ-OOK to NRZ-OOK and RZ-DPSK to NRZ-DPSK format conversion in a silicon microring resonator. *Opt. Express* **2012**, *20*, 27263–27272. [CrossRef] [PubMed]
68. Ding, Y.; Hu, H.; Galili, M.; Xu, J.; Liu, L.; Pu, M.; Mulvad, H.C.H.; Oxenløwe, L.K.; Peucheret, C.; Jeppesen, P.; et al. Generation of a 640 Gbit/s NRZ OTDM signal using a silicon microring resonator. *Opt. Express* **2011**, *19*, 6471–6477. [PubMed]
69. Zheng, Y.; Pu, M.; Yi, A.; Ou, X.; Ou, H. 4H-SiC microring resonators for nonlinear integrated photonics. *Opt. Lett.* **2019**, *44*, 5784–5787.
70. Shi, X.; Fan, W.; Yi, A.; Ou, X.; Rottwitt, K.; Ou, H. Dual-pump Optical Parametric Oscillation in a 4H-SiC-on-insulator Microring Resonator. In Proceedings of the European Conference on Lasers and Electro-Optics, Munich, Germany, 21–25 June 2021; Optica Publishing Group: Washington, DC, USA, 2021; p. cd_p_22.
71. Guidry, M.A.; Lukin, D.M.; Yang, K.Y.; Trivedi, R.; Vučković, J. Quantum optics of soliton microcombs. *Nat. Photonics* **2022**, *16*, 52–58.
72. Fortier, T.; Baumann, E. 20 years of developments in optical frequency comb technology and applications. *Commun. Phys.* **2019**, *2*, 153.
73. Fan, T.; Wu, X.; Vangapandu, S.R.; Hosseinnia, A.H.; Eftekhar, A.A.; Adibi, A. Racetrack microresonator based electro-optic phase shifters on a 3C silicon-carbide-on-insulator platform. *Opt. Lett.* **2021**, *46*, 2135–2138.
74. Powell, K.; Li, L.; Shams-Ansari, A.; Wang, J.; Meng, D.; Sinclair, N.; Deng, J.; Lončar, M.; Yi, X. Integrated silicon carbide electro-optic modulator. *Nat. Commun.* **2022**, *13*, 1851.
75. Cheng, C.H.; Fu, C.S.; Set, S.Y.; Yamashita, S.; Lin, G.R. All-Optical 25-Gbps Silicon Carbide Kerr Switching Intensity Modulator On Silicon Platform. In Proceedings of the 2022 Conference on Lasers and Electro-Optics (CLEO), San Jose, CA, USA, 15–20 May 2022; pp. 1–2.
76. Pelton, M. Modified spontaneous emission in nanophotonic structures. *Nat. Photonics* **2015**, *9*, 427–435.
77. Song, B.S.; Noda, S.; Asano, T.; Akahane, Y. Ultra-high-Q photonic double-heterostructure nanocavity. *Nat. Mater.* **2005**, *4*, 207–210.
78. Stoldt, Conrad ad Carraro, C.; Ashurt, R.; Gao, D.; Howe, R.; Maboudian, R. A low-temperature CVD process for silicon carbide MEMS. *Sens. Actuators A Phys.* **2002**, *97–98*, 410–415.
79. Habuka, H.; Ando, Y.; Tsuji, M. Room temperature process for chemical vapor deposition of amorphous silicon carbide thin film using monomethylsilane gas. *Surf. Coat. Technol.* **2011**, *206*, 1503–1506.
80. Wang, L. Dependence of chemical composition and bonding of amorphous SiC on deposition temperature and the choice of substrate. *J. -Non-Cryst. Solids* **2011**, *357*, 1063–1069.
81. Habuka, H.; Tsuji, M. Chemical vapor deposition of amorphous silicon carbide thin films on metal surfaces using monomethylsilane gas at low temperatures. *Surf. Coatings Technol.* **2013**, *217*, 88–93.
82. Yoshino, M.; Shimozuma, M.; Date, H.; Itoh, H.; Tagashira, H. Deposition of SiC films by ion-enhanced plasma chemical vapor deposition using tetramethylsilane+ H2. *Thin Solid Film.* **2005**, *492*, 207–211.
83. Frischmuth, T.; Schneider, M.; Radovic, I.; Siketic, Z.; Maurer, D.; Grille, T.; Shmid, U. Low temperature deposition of a-SiC:H thin films applying a dual plasma source process. *Thin Solid Film.* **2016**, *616*, 164–171.
84. Kwon, S.; Park, Y.; Ban, W.; Youn, C.; Lee, S.; Yang, J.; Jung, D.; Choi, T. Effect of plasma power on properties of hydrogenated amorphous silicon carbide hardmask films deposited by PECVD. *Vacuum* **2020**, *174*, 109187.
85. Jung, C.; Lim, D.; Jee, H.; Park, M.; Ku, S.; Yu, K.; Hong, B.; Lee, S.; Boo, J. Hydrogenated amorphous and crystalline SiC thin films grown by RF-PECVD and thermal MOCVD; comparative study of structural and optical properties. *Surf. Coat. Technol.* **2003**, *171*, 46–50.

86. Chen, E.; Du, G.; Zhang, Y.; Qin, X.; Lai, H.; Shi, W. RF-PECVD deposition and optical properties of hydrogenated amorphous silicon carbide thin films. *Ceram. Int.* **2014**, *40*, 9791–9797.
87. Hoshide, Y.; Komura, Y.; Tabata, A.; Kitagawa, A.; Kondo, A. Importance of H2 gas for growth of hot-wire CVD nanocrystalline 3C-SiC from SiH4/CH4/H2. *Thin Solid Film.* **2009**, *517*, 3520–3523. [CrossRef]
88. Tabata, A.; Komura, Y.; Narita, T.; Kondo, A. Growth of silicon carbide thin films by hot-wire chemical vapor deposition from SiH4/CH4/H2. *Thin Solid Film.* **2009**, *517*, 3516–3519. [CrossRef]
89. Klein, S.; Carius, R.; Finger, F.; Houben, L. Low substrate temperature deposition of crystalline SiC using HWCVD. *Thin Solid Film.* **2006**, *501*, 169–172. [CrossRef]
90. Finger, F.; Astakhov, O.; Bronger, T.; Carius, R.; Chen, T.; Dasgupta, A.; Gordijn, A.; Houben, L.; Huang, Y.; Klein, S.; et al. Microcrystalline silicon carbide alloys prepared with HWCVD as highly transparent and conductive window layers for thin film solar cells. *Thin Solid Film.* **2009**, *517*, 3507–3512. [CrossRef]
91. Zeiler, E.; Schwars, S.; Rosiwal, S.; Singer, R. Structural changes of tungsten heating filaments during CVD of diamond. *Mater. Sci. Eng. A* **2002**, *355*, 236–245. [CrossRef]
92. Tabata, A.; Naito, A. Structural changes in tungsten wire and their effect on the properties of hydrogenated nanocrystalline cubic silicon carbide thin films. *Thin Solid Film.* **2011**, *519*, 4451–4454. [CrossRef]
93. Komura, Y.; Tabata, A.; Narita, T.; Kondo, A. Influence of gas pressure on low-temperature preparation and film properties of nanocrystalline 3C-SiC thin films by HW-CVD using SiH4/CH4/H2 system. *Thin Solid Film.* **2008**, *516*, 633–636. [CrossRef]
94. Zhao, Q.; Li, J.; Zhou, H.; Wang, H.; Wang, B.; Yan, H. Parameters determining crystallinity in β-SiC thin films prepared by catalytic chemical vapor deposition. *J. Cryst. Growth* **2004**, *260*, 176–180. [CrossRef]
95. Swain, B.; Dusane, R. Multiphase structure of hydrogen diluted a-SiC:H deposited by HWCVD. *Mater. Chem. Phys.* **2006**, *99*, 240–246. [CrossRef]
96. Mahmood, A.; Muhl, S.; Machorro, R.; Lousa, A.; Esteve, J.; Heiras, J. Surface and optical analysis of SiCx films prepared by RF-RMS technique. *Diam. Relat. Mater.* **2006**, *15*, 71–79. [CrossRef]
97. Tang, H.; Tan, S.; Huang, Z.; Dong, S.; Jiang, D. Surface morphology of α-SIC coatings deposited by RF magnetron sputtering. *Surf. Coat. Technol.* **2005**, *197*, 161–167. [CrossRef]
98. Manivannan, R.; Sundararaj, S.; Dheenasagar, R.; Giridharan, K.; Sivaraman, P.; Udhayarani, V. Influence of Al2O3, SiC and B4C covalent multilayer PVD coating on surface properties of HSS rod. *Mater. Today Proc.* **2021**, *39*, 700–707. [CrossRef]
99. Liu, M.; Yang, Y.; Mao, Q.; Wei, Y.; Li, Y.; Ma, N.; Liu, H.; Liu, X.; Huang, Z. Influence of radio frequency magnetron sputtering parameters on the structure and performance of SiC films. *Ceram. Int.* **2021**, *47*, 24098–24105. [CrossRef]
100. Tavsanoglu, T.; Zayim, E.; Agirseven, O.; Yildirim, S.; Yucel, O. Optical, electrical and microstructural properties of SiC thin films deposited by reactive dc magnetron sputtering. *Thin Solid Film.* **2019**, *674*, 1–6. [CrossRef]
101. Gou, L.; Qi, C.; Ran, J.; Zheng, C. SiC film deposition by DC magnetron sputtering. *Thin Solid Film.* **1999**, *345*, 42–44. [CrossRef]
102. Knotek, O.; Löffler, F.; Wolkers, L. Amorphous SiC PVD coatings. *Diam. Relat. Mater.* **1993**, *2*, 528–530. [CrossRef]
103. Cheng, Y.; Huang, X.; Du, Z.; Xiao, J. Effect of sputtering power on the structure and optical band gap of SiC thin films. *Opt. Mater.* **2017**, *73*, 723–728. [CrossRef]
104. Seo, J.K.; Joung, Y.H.; Park, Y.; Won-Seok, C. Substrate temperature effect on the SiC passivation layer synthesized by an RF magnetron sputtering method. *Thin Solid Film.* **2011**, *519*, 6654–6657. [CrossRef]
105. Wang, Y.; Tuofu, Z.M.; Yue, Z.; Sun, H.; Gong, J.; Gao, J. Research on adhesion strength and optical properties of SiC films obtained via RF magnetron sputtering. *Chin. J. Phys.* **2020**, *64*, 79–86. [CrossRef]
106. Tang, H.; Tan, S.; Huang, Z. SiC Coatings Deposited by RF Magnetron Sputtering. *Key Eng. Mater.* **2007**, *280–283*, 1309–1312. [CrossRef]
107. Akshara, P.; Rajaram, G.; Krishna, G. Single composite target magnetron sputter deposition of crystalline and amorphous SiC thin films. *Mater. Res. Express* **2018**, *5*, 036410. [CrossRef]
108. Robertson, J. Diamond-like amorphous carbon. *Mater. Sci. Eng. R Rep.* **2002**, *37*, 129–281. [CrossRef]
109. Zielinski, M.; S Ndiaye, T.C.; Juillaguet, S.; Lewandowska, R.; Portail, M.; Leycuras, A.; Camassel, J. Strain and wafer curvature of 3C-SiC films on silicon: Influence of the growth conditions. *Phys. Status Solidi A* **2007**, *204*, 981–986. [CrossRef]
110. Cimalla, V.; Pezoldt, J.; Ambacherand, O. Group III nitride and SiC based MEMS and NEMS: Materials properties, technology and applications. *J. Phys. D Appl. Phys.* **2007**, *40*, 6386.
111. Zimbone, M.; Mauceri, M.; Litrico, G.; Barbagiovanni, E.G.; Bongiorno, C.; La Via, F. Protrusions reduction in 3C-SiC thin film on Si. *J. Cryst. Growth* **2018**, *498*, 248–257. [CrossRef]
112. Liaw, P.; Davis, R. Epitaxial growth and characterization of β-SiC thin films. *J. Electrochem. Soc.* **1985**, *132*, 642. [CrossRef]
113. Zimbone, M.; Sarikov, A.; Bongiorno, C.; Marzegalli, A.; Scuderi, V.; Calabretta, C.; Miglio, L.; La Via, F. Extended defects in 3C-SiC: Stacking faults, threading partial dislocations, and inverted domain boundaries. *Acta Mater.* **2021**, *213*, 116915. [CrossRef]
114. Nagasawa, H.; Yagi, K.; Kawahara, T.; Hatta, N. Reducing planar defects in 3C–SiC. *Chem. Vap. Depos.* **2006**, *12*, 502–508. [CrossRef]
115. Shibahara, K.; Nishino, S.; Matsunami, H. Antiphase-domain-free growth of cubic SiC on Si (100). *Appl. Phys. Lett.* **1987**, *50*, 1888–1890. [CrossRef]
116. La Via, F.; Camarda, M.; Canino, A.; Severino, A.; LA MAGNA, A.; Mauceri, M.; Vecchio, C.; Crippa, D. Fast growth rate epitaxy by chloride precursors. *Mater. Sci. Forum* **2013**, *740*, 167–172. [CrossRef]

117. Chassagne, T.; Ferro, G.; Chaussende, D.; Cauwet, F.; Monteil, Y.; Bouix, J. A comprehensive study of SiC growth processes in a VPE reactor. *Thin Solid Film.* **2002**, *402*, 83–89. [CrossRef]
118. Nishino, S.; Powell, J.A.; Will, H.A. Production of large-area single-crystal wafers of cubic SiC for semiconductor devices. *Appl. Phys. Lett.* **1983**, *42*, 460–462. [CrossRef]
119. Björketun, L.O.; Hultman, L.; Ivanov, I.; Wahab, Q.; Sundgren, J.E. Interfacial void formation during vapor phase growth of 3C-SiC on Si (0 0 1) and Si (1 1 1) substrates. Characterization by transmission electron microscopy. *J. Cryst. Growth* **1997**, *182*, 379–388. [CrossRef]
120. Anzalone, R.; Litrico, G.; Piluso, N.; Reitano, R.; Alberti, A.; Fiorenza, P.; Coffa, S.; La Via, F. Carbonization and transition layer effects on 3C-SiC film residual stress. *J. Cryst. Growth* **2017**, *473*, 11–19. [CrossRef]
121. Nagasawa, H.; Yamaguchi, Y. Suppression of etch pit and hillock formation on carbonization of Si substrate and low temperature growth of SiC. *J. Cryst. Growth* **1991**, *115*, 612–616. [CrossRef]
122. Kordina, O.; Björketun, L.O.; Henry, A.; Hallin, C.; Glass, R.; Hultman, L.; Sundgren, J.E.; Janzen, E. Growth of 3CSiC on on-axis Si (100) substrates by chemical vapor deposition. *J. Cryst. Growth* **1995**, *154*, 303–314. [CrossRef]
123. Portail, M.; Zielinski, M.; Chassagne, T.; Roy, S.; Nemoz, M. Comparative study of the role of the nucleation stage on the final crystalline quality of (111) and (100) silicon carbide films deposited on silicon substrates. *J. Appl. Phys.* **2009**, *105*, 083505. [CrossRef]
124. Anzalone, R.; Frazzeto, N.; La Via, F. Apparatus for growing a semiconductor wafer and associated manufacturing process, 2021.
125. Anzalone, R.; Zimbone, M.; Calabretta, C.; Mauceri, M.; Alberti, A.; Reitano, R.; La Via, F. Temperature Investigation on 3C-SiC Homo-Epitaxy on Four-Inch Wafers. *Materials* **2019**, *12*, 3293. [CrossRef]
126. Anzalone, R.; Camarda, M.; Locke, C.; Carballo, J.; Piluso, N.; La Magna, A.; Volinsky, A.A.; Saddow, S.E.; La Via, F. Stress nature investigation on heteroepitaxial 3C–SiC film on (100) Si substrates. *J. Mater. Res.* **2013**, *28*, 129–135. [CrossRef]
127. Kollmuss, M.; Köhler, J.; Ou, H.Y.; Fan, W.C.; Chaussende, D.; Hock, R.; Wellmann, P.J. Chemical Vapor Deposition of 3C-SiC on [100] Oriented Silicon at Low Temperature< 1200° C for Photonic Applications. *Mater. Sci. Forum* **2022**, *1062*, 119–124.
128. Wilhelm, M.; Rieth, M.; Brandl, M.; Wibowo, R.A.; Hock, R.; Wellmann, P. Optimization of growth parameters for growth of high quality heteroepitaxial 3C–SiC films at 1200 C. *Thin Solid Film.* **2015**, *577*, 88–93. [CrossRef]
129. Gao, Y.; Edgar, J.; Chaudhuri, J.; Cheema, S.; Sidorov, M.; Braski, D. Low-temperature chemical-vapor deposition of 3C–SiC films on Si (1 0 0) using SiH4–C2H4–HCl–H2. *J. Cryst. Growth* **1998**, *191*, 439–445. [CrossRef]
130. Fissel, A.; Schröter, B.; Kaiser, U.; Richter, W. Advances in the molecular-beam epitaxial growth of artificially layered heteropolytypic structures of SiC. *Appl. Phys. Lett.* **2000**, *77*, 2418–2420. [CrossRef]
131. Kimoto, T.; Itoh, A.; Matsunami, H. Step-controlled epitaxial growth of high-quality SiC layers. *Phys. Status Solidi B* **1997**, *202*, 247–262. [CrossRef]
132. Matsunami, H. Technological breakthroughs in growth control of silicon carbide for high power electronic devices. *Jpn. J. Appl. Phys.* **2004**, *43*, 6835. [CrossRef]
133. Powell, J.; Larkin, D. Process-induced morphological defects in epitaxial CVD silicon carbide. *Phys. Status Solidi B* **1997**, *202*, 529–548. [CrossRef]
134. Jokubavicius, V.; Yazdi, G.R.; Liljedahl, R.; Ivanov, I.G.; Yakimova, R.; Syväjärvi, M. Lateral enlargement growth mechanism of 3C-SiC on off-oriented 4H-SiC substrates. *Cryst. Growth Des.* **2014**, *14*, 6514–6520. [CrossRef]
135. Jokubavicius, V.; Yazdi, G.R.; Liljedahl, R.; Ivanov, I.G.; Sun, J.; Liu, X.; Schuh, P.; Wilhelm, M.; Wellmann, P.; Yakimova, R. Single domain 3C-SiC growth on off-oriented 4H-SiC substrates. *Cryst. Growth Des.* **2015**, *15*, 2940–2947. [CrossRef]
136. Shi, Y.; Jokubavicius, V.; Höjer, P.; Ivanov, I.G.; Yazdi, G.R.; Yakimova, R.; Syväjärvi, M.; Sun, J. A comparative study of high-quality C-face and Si-face 3C-SiC (1 1 1) grown on off-oriented 4H-SiC substrates. *J. Phys. D Appl. Phys.* **2019**, *52*, 345103. [CrossRef]
137. Vasiliauskas, R.; Marinova, M.; Hens, P.; Wellmann, P.; Syvajarvi, M.; Yakimova, R. Nucleation control of cubic silicon carbide on 6H-substrates. *Cryst. Growth Des.* **2012**, *12*, 197–204. [CrossRef]
138. Vasiliauskas, R.; Marinova, M.; Syväjärvi, M.; Mantzari, A.; Andreadou, A.; Polychroniadis, E.K.; Yakimova, R. Sublimation epitaxy of cubic silicon carbide in vacuum. *J. Phys. Conf. Ser.* **2010**, *223*, 012014. [CrossRef]
139. Neudeck, P.G.; Trunek, A.J.; Spry, D.J.; Powell, J.A.; Du, H.; Skowronski, M.; Huang, X.R.; Dudley, M. CVD Growth of 3C-SiC on 4H/6H Mesas. *Chem. Vap. Depos.* **2006**, *12*, 531–540. [CrossRef]
140. Schuh, P.; Vecera, P.; Hirsch, A.; Syväjärvi, M.; Litrico, G.; La Via, F.; Mauceri, M.; Wellmann, P.J. Physical vapor growth of double position boundary free, quasi-bulk 3C-SiC on high quality 3C-SiC on Si CVD templates. *Mater. Sci. Forum* **2016**, *858*, 89–92. [CrossRef]
141. Wellmann, P.J.; Schuh, P.; Kollmuss, M.; Schöler, M.; Steiner, J.; Zielinski, M.; Mauceri, M.; La Via, F. Prospects of Bulk Growth of 3C-SiC Using Sublimation Growth. *Mater. Sci. Forum* **2020**, *1004*, 113–119. [CrossRef]
142. Wellmann, P.; Ohtani, N.; Rupp, R. *Wide Bandgap Semiconductors for Power Electronics: Materials, Devices, Applications*; John Wiley & Sons: Weinheim, Germany, 2021.
143. Schuh, P.; La Via, F.; Mauceri, M.; Zielinski, M.; Wellmann, P.J. Growth of large-area, stress-free, and bulk-like 3C-SiC (100) using 3C-SiC-on-Si in vapor phase growth. *Materials* **2019**, *12*, 2179. [CrossRef]
144. Schuh, P.; Steiner, J.; La Via, F.; Mauceri, M.; Zielinski, M.; Wellmann, P.J. Limitations during vapor phase growth of bulk (100) 3C-SiC using 3C-SiC-on-SiC seeding stacks. *Materials* **2019**, *12*, 2353. [CrossRef]

145. La Via, F.; Zimbone, M.; Bongiorno, C.; La Magna, A.; Fisicaro, G.; Deretzis, I.; Scuderi, V.; Calabretta, C.; Giannazzo, F.; Zielinski, M. New Approaches and Understandings in the Growth of Cubic Silicon Carbide. *Materials* **2021**, *14*, 5348. [CrossRef]
146. Dubecký, F.; Gombia, E.; Ferrari, C.; Zat'ko, B.; Vanko, G.; Baldini, M.; Kováč, J.; Baček, D.; Kováč, P.; Hrkút, P.; et al. Characterization of epitaxial 4H-SiC for photon detectors. *J. Instrum.* **2012**, *7*, P09005. [CrossRef]
147. Schoeler, M. Characterization of protrusions and stacking faults in 3C-SiC grown by sublimation epitaxy using 3C-SiC-on-Si seeding layers. *Adv. Mater. Proc.* **2017**, *2*, 774–778.
148. Kollmuss, M.; Schöler, M.; Anzalone, R.; Mauceri, M.; La Via, F.; Wellmann, P.J. Large Area Growth of Cubic Silicon Carbide Using Close Space PVT by Application of Homoepitaxial Seeding. *Mater. Sci. Forum* **2022**, *1062*, 74–78. [CrossRef]
149. Schöler, M.; La Via, F.; Mauceri, M.; Wellmann, P. Overgrowth of Protrusion Defects during Sublimation Growth of Cubic Silicon Carbide Using Free-Standing Cubic Silicon Carbide Substrates. *Cryst. Growth Des.* **2021**, *21*, 4046–4054. [CrossRef]
150. Schuh, P.; Künecke, U.; Litrico, G.; Mauceri, M.; La Via, F.; Monnoye, S.; Zielinski, M.; Wellmann, P.J. Vapor Growth of 3C-SiC Using the Transition Layer of 3C-SiC on Si CVD Templates. *Mater. Sci. Forum* **2019**, *963*, 149–152. [CrossRef]
151. Tairov, Y.M.; Tsvetkov, V.F. Investigation of growth processes of ingots of silicon carbide single crystals. *J. Cryst. Growth* **1978**, *43*, 209–212. [CrossRef]
152. Scace, R.I.; Slack, G.A. Solubility of Carbon in Silicon and Germanium. *J. Chem. Phys.* **1959**, *30*, 1551–1555. [CrossRef]
153. Harada, S.; Yamamoto, Y.; Xiao, S.; Koike, D.; Mutoh, T.; Murayama, K.; Aoyagi, K.; Sakai, T.; Tagawa, M.; Ujihara, T. Dislocation Conversion During SiC Solution Growth for High-Quality Crystals. *Mater. Sci. Forum* **2015**, *821–823*, 3–8. [CrossRef]
154. Wellmann, P.J. Power Electronic Semiconductor Materials for Automotive and Energy Saving Applications—SiC, GaN, Ga2O3, and Diamond. *Z. Anorg. Allg. Chem.* **2017**, *643*, 1312–1322. [CrossRef]
155. Wellmann, P.; Straubinger, T.; Desperrier, P.; Mueller, R.; ünecke, U.; Sakwe, S.A.; Schmitt, H.; Winnacker, A.; Blanquet, E.; Dedulle, J.M.; et al. Modified Physical Vapor Transport Growth of SiC—Control of Gas Phase Composition for Improved Process Conditions. *Mat. Sci. Forum* **2005**, *483–485*, 25–30. [CrossRef]
156. Chaussende, D.; Ohtani, N. Silicon carbide. In *Single Crystals of Electronic Materials*; Elsevier: Amsterdam, The Netherlands, 2019; pp. 129–179.
157. Wellmann, P.J. Review of SiC crystal growth technology. *Semicond. Sci. Technol.* **2018**, *33*, 103001. [CrossRef]
158. Koga, K.; Fujikawa, Y.; Ueda, Y.; Yamaguchi, T. Growth and characterization of 6H-SiC bulk crystals by the sublimation method. In *Amorphous and Crystalline Silicon Carbide IV*; Springer: Berlin/Heidelberg, Germany, 1992; pp. 96–100.
159. Neudeck, P.G.; Powell, J.A. Performance limiting micropipe defects in silicon carbide wafers. *IEEE Electron Device Lett.* **1994**, *15*, 63–65. [CrossRef]
160. Agarwal, A.; Fatima, H.; Haney, S.; Ryu, S.H. A New Degradation Mechanism in High-Voltage SiC Power MOSFETs. *IEEE Electron Device Lett.* **2007**, *28*, 587–589. [CrossRef]
161. Veliadis, V.; Hearne, H.; Stewart, E.J.; Snook, M.; Chang, W.; Caldwell, J.D.; Ha, H.C.; El-Hinnawy, N.; Borodulin, P.; Howell, R.S.; et al. Degradation and Full Recovery in High-Voltage Implanted-Gate SiC JFETs Subjected to Bipolar Current Stress. *IEEE Electron Device Lett.* **2012**, *33*, 952–954. [CrossRef]
162. Chen, H.; Raghothamachar, B.; Vetter, W.; Dudley, M.; Wang, Y.; Skromme, B.J. Effects of Different Defect Types on the Performance of Devices Fabricated on a 4H-SiC Homoepitaxial Layer. *MRS Proc.* **2006**, *911*. [CrossRef]
163. Sakwe, S.; Jang, Y.S.; Wellmann, P. Defect Etching of SiC on non-polar faces. *Mater. Sci. Forum* **2007**, *556–557*, 243–246. [CrossRef]
164. Katsuno, M.; Ohtani, N.; Takahashi, J.; Yashiro, H.; Kanaya, M. Mechanism of Molten KOH Etching of SiC Single Crystals: Comparative Study with Thermal Oxidation. *Jpn. J. Appl. Phys.* **1999**, *38*, 4661–4665. [CrossRef]
165. Vetter, W.M.; Dudley, M. Micropipes and the closure of axial screw dislocation cores in silicon carbide crystals. *J. Appl. Phys.* **2004**, *96*, 348–353. [CrossRef]
166. Vetter, W.M.; Dudley, M. Micropipes in Silicon Carbide Crystals: Do all Screw Dislocations have Open Cores? *J. Mater. Res.* **2011**, *15*, 1649–1652. [CrossRef]
167. Chen, P.C.; Miao, W.C.; Ahmed, T.; Pan, Y.Y.; Lin, C.L.; Chen, S.C.; Kuo, H.C.; Tsui, B.Y.; Lien, D.H. Defect Inspection Techniques in SiC. *Nanoscale Res. Lett.* **2022**, *17*, 30. [CrossRef]
168. Dudley, M.; Huang, X.R.; Huang, W.; Powell, A.; Wang, S.; Neudeck, P.; Skowronski, M. The mechanism of micropipe nucleation at inclusions in silicon carbide. *Appl. Phys. Lett.* **1999**, *75*, 784–786. [CrossRef]
169. Hofmann, D.; Schmitt, E.; Bickermann, M.; Kölbl, M.; Wellmann, P.J.; Winnacker, A. Analysis on defect generation during the SiC bulk growth process. *Mater. Sci. Eng.* **1999**, *61–62*, 48–53. [CrossRef]
170. Kuhr, T.A.; Sanchez, E.K.; Skowronski, M.; Vetter, W.M.; Dudley, M. Hexagonal voids and the formation of micropipes during SiC sublimation growth. *J. Appl. Phys.* **2001**, *89*, 4625–4630. [CrossRef]
171. Sanchez, E.K.; Kuhr, T.; Heydemann, V.D.; Snyder, D.W.; Rohrer, G.S.; Skowronski, M. Formation of thermal decomposition cavities in physical vapor transport of silicon carbide. *J. Electron. Mater.* **2000**, *29*, 347–352. [CrossRef]
172. Powell, A.R.; Sumakeris, J.J.; Khlebnikov, Y.; Paisley, M.J.; Leonard, R.T.; Deyneka, E.; Gangwal, S.; Ambati, J.; Tsevtkov, V.; Seaman, J.; et al. Bulk Growth of Large Area SiC Crystals. *Mater. Sci. Forum* **2016**, *858*, 5–10. [CrossRef]
173. Sanchez, E.K.; Liu, J.Q.; De Graef, M.; Skowronski, M.; Vetter, W.M.; Dudley, M. Nucleation of threading dislocations in sublimation grown silicon carbide. *J. Appl. Phys.* **2002**, *91*, 1143–1148. [CrossRef]
174. Pirouz, P.; Zhang, M.; Demenet, J.L.; Hobgood, H.M. Transition from brittleness to ductility in SiC. *J. Phys. Condens. Matter* **2002**, *14*, 12929–12945. [CrossRef]

175. Ohtani, N.; Katsuno, M.; Fujimoto, T.; Nakabayashi, M.; Tsuge, H.; Yashiro, H.; Aigo, T.; Hirano, H.; Hoshino, T.; Ohashi, W. Analysis of Basal Plane Bending and Basal Plane Dislocations in 4H-SiC Single Crystals. *Jpn. J. Appl. Phys.* **2009**, *48*, 065503. [CrossRef]
176. Shioura, K.; Shinagawa, N.; Izawa, T.; Ohtani, N. Structural characterization of the grown crystal/seed interface of physical vapor transport grown 4H-SiC crystals using Raman microscopy and x-ray topography. *J. Cryst. Growth* **2019**, *515*, 58–65. [CrossRef]
177. Gao, B.; Kakimoto, K. Three-Dimensional Modeling of Basal Plane Dislocations in 4H-SiC Single Crystals Grown by the Physical Vapor Transport Method. *Cryst. Growth Des.* **2014**, *14*, 1272–1278. [CrossRef]
178. Nakano, T.; Shinagawa, N.; Yabu, M.; Ohtani, N. Formation and multiplication of basal plane dislocations during physical vapor transport growth of 4H-SiC crystals. *J. Cryst. Growth* **2019**, *516*, 51–56. [CrossRef]
179. Stockmeier, M.; Muller, R.; Sakwe, S.A.; Wellmann, P.J.; Magerl, A. On the lattice parameters of silicon carbide. *J. Appl. Phys.* **2009**, *105*, 033511. [CrossRef]
180. Steiner, J.; Wellmann, P.J. Impact of Mechanical Stress and Nitrogen Doping on the Defect Distribution in the Initial Stage of the 4H-SiC PVT Growth Process. *Materials* **2022**, *15*, 1897. [CrossRef]
181. Ellefsen, O.M.; Arzig, M.; Steiner, J.; Wellmann, P.; Runde, P. Optimization of the SiC Powder Source Material for Improved Process Conditions During PVT Growth of SiC Boules. *Materials* **2019**, *12*, 3272. [CrossRef]
182. Steiner, J.; Arzig, M.; Hsiao, T.C.; Wellmann, P.J. Optimization of the SiC Powder Source Size Distribution for the Sublimation Growth of Long Crystals Boules. *Mater. Sci. Forum* **2019**, *963*, 42–45. [CrossRef]
183. Wellmann, P.; Bickermann, M.; Hofmann, D.; Kadinski, L.; Selder, M.; Straubinger, T.; Winnacker, A. In situ visualization and analysis of silicon carbide physical vapor transport growth using digital X-ray imaging. *J. Cryst. Growth* **2000**, *216*, 263–272. [CrossRef]
184. Arzig, M.; Salamon, M.; Uhlmann, N.; Wellmann, P.J. Investigation of the Growth Kinetics of SiC Crystals during Physical Vapor Transport Growth by the Application of In Situ 3D Computed Tomography Visualization. *Adv. Eng. Mater.* **2019**, *22*. [CrossRef]
185. Arzig, M.; Steiner, J.; Salamon, M.; Uhlmann, N.; Wellmann, P.J. Influence of Morphological Changes in a Source Material on the Growth Interface of 4h-Sic Single Crystals. *Materials* **2019**, *12*, 2591. [CrossRef] [PubMed]
186. Arzig, M.; Salamon, M.; Hsiao, T.C.; Uhlmann, N.; Wellmann, P.J. Influence of the growth interface shape on the defect characteristics in the facet region of 4H-SiC single crystals. *J. Cryst. Growth* **2020**, *532*, 125436. [CrossRef]
187. Kitanin, E.L.; Ramm, M.S.; Ris, V.V.; Schmidt, A.A. Heat transfer through source powder in sublimation growth of SiC crystal. *Mater. Sci. Eng. B* **1998**, *55*, 174–183. [CrossRef]
188. Ma, R.H.; Zhang, H.; Dudley, M.; Prasad, V. Thermal system design and dislocation reduction for growth of wide band gap crystals. *J. Cryst. Growth* **2003**, *258*, 318–330. [CrossRef]
189. Selder, M.; Kadinski, L.; Makarov, Y.; Durst, F.; Wellmann, P.; Straubinger, T.; Hofmann, D.; Karpov, S.; Ramm, M. Global numerical simulation of heat and mass transfer for SiC bulk crystal growth by PVT. *J. Cryst. Growth* **2000**, *211*, 333–338. [CrossRef]
190. Wellmann, P.; Herro, Z.; Selder, M.; Durst, F.; Püsche, R.; Hundhausen, M.; Ley, L.; Winnacker, A. Investigation of mass transport during SiC PVT growth using digital X-ray imaging, 13C labeling of source material and numerical modeling. *Mat. Sci. Forum* **2003**, *433–436*, 9–12. [CrossRef]
191. Ma, R.H.; Zhang, H.; Prasad, V.; Dudley, M. Growth kinetics and thermal stress in the sublimation growth of silicon carbide. *Cryst. Growth Des.* **2002**, *2*, 213–220. [CrossRef]
192. Selder, M.; Kadinski, L.; Durst, F.; Straubinger, T.; Wellmann, P.; Hofmann, D. Numerical simulation of thermal stress formation during PVT-growth of SiC bulk crystals. *Mat. Sci. Forum* **2001**, *353–356*, 65–68. [CrossRef]
193. Gallien, B.; Albaric, M.; Duffar, T.; Kakimoto, K.; M'Hamdi, M. Study on the usage of a commercial software (Comsol-Multiphysics) for dislocation multiplication model. *J. Cryst. Growth* **2017**, *457*, 60–64. [CrossRef]
194. Nguyen, B.D.; Rausch, A.M.; Steiner, J.; Wellmann, P.; Sandfeld, S. On the importance of dislocation flow in continuum plasticity models for semiconductor materials. *J. Cryst. Growth* **2020**, *532*, 125414. [CrossRef]
195. Matsunami, H.; Kimoto, T. Step-controlled epitaxial growth of SiC: High quality homoepitaxy. *Mater. Sci. Eng. R-Rep.* **1997**, *20*, 125–166. [CrossRef]
196. Burton, W.; Cabrera, N.; Frank, F. The growth of crystals and the equilibrium structure of their surfaces. *Philos. Trans. R. Soc. Lond. Ser. A Math. Phys. Sci.* **1951**, *243*, 299–358.
197. Jepps, N.W.; Page, T.F. Polytypic transformations in silicon carbide. *Prog. Cryst. Growth Charact.* **1983**, *7*, 259–307. [CrossRef]
198. Kallinger, B.; Polster, S.; Berwian, P.; Friedrich, J.; Danilewsky, A.N. Experimental verification of the model by Klapper for 4H-SiC homoepitaxy on vicinal substrates. *J. Appl. Phys.* **2013**, *114*, 183507. [CrossRef]
199. Kallinger, B.; Thomas, B.; Polster, S.; Berwian, P.; Friedrich, J. Dislocation Conversion and Propagation during Homoepitaxial Growth of 4H-SiC. *Mater. Sci. Forum* **2010**, *645–648*, 299–302. [CrossRef]
200. Ou, H.; Ou, Y.; Argyraki, A.; Schimmel, S.; Kaiser, M.; Wellmann, P.; Linnarsson, M.K.; Jokubavicius, V.; Sun, J.; Liljedahl, R.; et al. Advances in wide bandgap SiC for optoelectronics. *Eur. Phys. J. B* **2014**, *87*, 58. [CrossRef]

201. Lu, W.; Ou, Y.; Fiordaliso, E.M.; Iwasa, Y.; Jokubavicius, V.; Syväjärvi, M.; Kamiyama, S.; Petersen, P.M.; Ou, H. White light emission from fluorescent SiC with porous surface. *Sci. Rep.* **2017**, *7*, 9798. [CrossRef] [PubMed]
202. Tarekegne, A.T.; Shi, X.; Gan, Y.; Chen, Y.; Ou, H. Dependence of Photoluminescence Emission on Excitation Power and Temperature in Highly Doped 6H-SiC. *Phys. Rev. Appl.* **2020**, *13*, 064002. [CrossRef]

MDPI AG
Grosspeteranlage 5
4052 Basel
Switzerland
Tel.: +41 61 683 77 34

Materials Editorial Office
E-mail: materials@mdpi.com
www.mdpi.com/journal/materials

www.ingramcontent.com/pod-product-compliance
Lightning Source LLC
Chambersburg PA
CBHW042010260926
51850CB00020B/2968
9783725822386